EVOLUTION IN PERSPECTIVE

Evolution in Perspective
The Science Teacher's Compendium

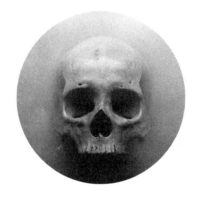

Rodger W. Bybee, Editor

An NSTA Press Journals Collection

NATIONAL SCIENCE TEACHERS ASSOCIATION
Arlington, Virginia

Claire Reinburg, Director
Andrew Cocke, Associate Editor
Judy Cusick, Associate Editor
Betty Smith, Associate Editor

The Science Teacher, Jennifer Henderson, Managing Editor
ART AND DESIGN Linda Olliver, Director
    Cover photo by Myron, courtesy of Getty Images.
NSTA WEB Tim Weber, Webmaster
PERIODICALS PUBLISHING Shelley Carey, Director
PRINTING AND PRODUCTION Catherine Lorrain, Director
    Nguyet Tran, Assistant Production Manager
    Jack Parker, Electronic Prepress Technician
PUBLICATIONS OPERATIONS Hank Janowsky, Manager
*sci*LINKS Tyson Brown, Manager
    David Anderson, Web and Development Coordinator

NATIONAL SCIENCE TEACHERS ASSOCIATION
Gerald F. Wheeler, Executive Director
David Beacom, Publisher

Copyright © 2004 by the National Science Teachers Association.
All rights reserved. Printed in the United States of America by Victor Graphics.

Evolution in Perspective: The Science Teacher's Compendium
    NSTA Stock Number: PB181X
08   07   06   5   4   3   2

Library of Congress Cataloging-in-Publication Data
Evolution in perspective : the science teacher's compendium/edited by
Rodger W. Bybee.
   p. cm.
 ISBN 0-87355-234-2
 1. Evolution (Biology)—Study and teaching (Secondary) I. Bybee,
Rodger W.
  QH362.E853 2003
  576.8—dc22
                       2003021609

NSTA is committed to publishing quality materials that promote the best in inquiry-based science education. However, conditions of actual use may vary and the safety procedures and practices described in this book are intended to serve only as a guide. Additional precautionary measures may be required. NSTA and the author(s) do not warrant or represent that the procedures and practices in this book meet any safety code or standard or federal, state, or local regulations. NSTA and the author(s) disclaim any liability for personal injury or damage to property arising out of or relating to the use of this book including any recommendations, instructions, or materials contained therein.

Permission is granted in advance for photocopying brief excerpts for one-time use in a classroom or workshop. Requests involving electronic reproduction should be directed to Permissions/NSTA Press, 1840 Wilson Blvd., Arlington, VA 22201-3000; fax 703-526-9754. Permissions requests for coursepacks, textbooks, and other commercial uses should be directed to Copyright Clearance Center, 222 Rosewood Dr., Danvers, MA 01923; fax 978-646-8600; *www.copyright.com*.

About the cover: The image on the cover is a photograph of a modern skull, age unknown. Photograph by Myron, courtesy of Getty Images.

CONTENTS

Foreword ix
"Turn on" the evolution lightbulb
Janet Gerking

Introduction xi
Evolution in perspective
Rodger W. Bybee

NSTA Position Statement xix
The teaching of evolution

SECTION 1: THE SCIENTIFIC PERSPECTIVE

Arguing for Evolution 1
Francisco J. Ayala
(February 2000)

Evidence for Evolution 5
G. Brent Dalrymple
(October 2000)

Thought Patterns in Science & Creationism 9
Understanding the vast difference between creationism and evolutionary theory
John A. Moore
(May 2000)

SECTION II THE EDUCATIONAL PERSPECTIVE

Evolution and the Nature of Science 15
A National Science Education Standards perspective
Rodger W. Bybee
(Special for this compendium)

Do Standards Matter? 25
How the quality of state standards relates to evolution instruction
Randy Moore
(January 2002)

Evolution: Don't Debate, Educate 29
Teach inquiry and the nature of science
Rodger W. Bybee
(October 2000)

It's Not Just a Theory 37
Why teachers need to address the nature of science and the "hidden" curriculum
DeWayne A. Backhus
(April 2002)

Evolution and Intelligent Design: 43
Understanding the issues surrounding evolution and intelligent design and dealing with the controversy
John R. Staver
(November 2003)

SECTION III THE SCIENCE TEACHER'S PERSPECTIVE

Attitudes Toward Evolution 49
Membership in professional organizations and standards use are associated with strong evolution teaching
Jeffrey Weld and Jill C. McNew
(December 1999)

Investigating Island Evolution 57
A Galapagos-based lesson using the 5E instructional model
Anthony V. DeFina
(February 2002)

Searching for the Perfect Lesson 63
Teaching evolution to a diverse biology class
Susan Stone Plati
(December 2001)

Comparing Common Origins 67
Using biotechnology to teach evolution
John McLaughlin and George Glasson
(November 2001)

Symbiosis: An Evolutionary Innovator 73
Blurring the concept of individuality, symbiosis tangles the phylogenetic trees
Emily Case
(April 2003)

A Teaching Guide to Evolution 79
Discovering evolution through molecular evidence
Thomas G. Gregg, Gary R. Janssen, and J.K. Bhattacharjee
(November 2003)

Appendix: List of Contributors 93

Index 95

FOREWORD

"Turn on" the evolution lightbulb

As a high school science teacher, I was always fascinated with my students' ability to "turn on" and "turn off" the material they wanted to learn. They were always selective in determining whether the information would be important to them one day and whether they wanted to spend time learning it. I never had to worry, however, about getting their attention when the word "evolution" appeared in the text, a video, or the activities. The majority of students immediately conjured up a vision of something forbidden, controversial, and downright wrong. I grew to expect their strong reactions to the word itself, their misconceptions about evolution and the word "theory," and their attentiveness during the learning process as we explored yearlong themes related to evolution and change. Some students changed their attitudes toward evolution as they inquired into the processes of change, but I also know that many students did not. Students always wrote at the end of the year about the evidence for evolution, which we had slowly and consistently developed throughout the course. I'll never forget one short essay that stated, "This was a lot of fun, but I don't believe any of it."

Teaching evolution does not just require aligning our curriculum to the National Science Education Standards (NSES) or putting the material in a textbook or unit on change and adaptation. Teaching evolution requires us to develop a whole way of thinking among our student population. Understanding the nature of science is fundamental to understanding evolution and understanding theories. We can build our case for evolution throughout the school year, but we cannot change the way our students think about evolution without focusing on the nature of science and working to dispel myths about the word "theory" itself.

Over the past five years, *The Science Teacher*, NSTA's secondary science journal, has reviewed and published original manuscripts that addressed this very issue. At first we concentrated on manuscripts demonstrating classroom activities for teachers to teach evolution. We were more concerned about student attitudes on the topic than we were about the science educators who taught the classes. But one detailed research study changed our focus on evolution, when Jeffrey Weld and Jill C. McNew submitted their unsolicited manuscript, "Attitudes toward Evolution," in the summer of 1999. Observing a

FOREWORD

"polar discord" among preservice teachers on the topic of creationism and evolution, the authors pursued a study to determine how much variation exists among teachers regarding this "controversial unifying concept" (Weld and McNew, December 1999, 27).

Their findings had national implications and revealed startling statistics. From Oklahoma, where they found 33 percent of science teachers placed little or no emphasis on evolution, to Illinois, where 30 percent believed both creationism and evolution should be taught in high school biology classes, statistics clearly demonstrated that national standards for teaching evolution theory were not being met.

While evolution is taught as an integral part of the science curriculum at the university level, secondary courses often are heavily influenced by personal views, and parental, student, and administrative pressure. Secondary textbooks, old and outdated in many classrooms, do not reflect current research and tend to place evolution in its own chapter, separate from all the concepts with which it intertwines. Secondary science courses, despite great strides in developing integrated curricula, usually still contain short, segmented units rather than an emphasis on overall concepts and themes central to science. Finding a textbook demonstrating the fundamental influence of evolution in all aspects of biological and physical science for the seventh grade, for instance—and then getting it adopted by the school district—is a huge challenge.

Textbook designers faced a real dilemma as well. While national standards cite evolution as a fundamental concept to the teaching of science, state standards began to erode the terminology and emphasis on evolution. Some states removed the word altogether, often referring to "adaptation" or "change."

Teachers were left in the middle of the debate on whether to teach to national or state science standards. Required to teach to state standards rather than national standards in many districts, teachers also faced a narrowing curriculum that emphasized facts and content rather than the overall picture of the nature of science. Pressure mounted in subsequent years to follow the textbook and not vary from the mandated curriculum, and the vicious cycle continued.

But the real significance of the Weld and McNew study lies in the background information they gathered. They found "no difference between the emphasis placed on evolution by males or females, new teachers versus veterans, rural versus urban and suburban teachers, or those with bachelor's vs. master's degrees" when teaching evolution in public school biology. Instead they found that the factors most important to identifying teacher's philosophies toward emphasizing evolution in their secondary classrooms are:

- Coursework and independent study in the nature or philosophy of science,
- Membership in professional science teacher organizations, and
- Use of national standards in guiding their practice (Weld and McNew, Dec. 1999, 31).

Indeed Weld and McNew's findings were significant and led us to emphasize the role of secondary science teachers in addressing evolution in terms of the nature and philosophy of science. Over the next several years, we published articles by scientists, professors, and classroom teachers who addressed evolution and its role in understanding the nature of science. Submitted manuscripts were peer reviewed and aligned with the objectives outlined in the *National Science Education Standards* (NRC, 1996), the American Association for the Advancement of Science (AAAS) *Benchmarks for Science Literacy* (1993), and NSTA's position statement *The Teaching of Evolution* (1997). This compendium is a result of that effort.

JANET L. GERKING
Field Editor
The Science Teacher

INTRODUCTION

EVOLUTION IN PERSPECTIVE

RODGER W. BYBEE

Science teachers have the opportunity to introduce students to one of the greatest intellectual achievements in history. I speak here, of course, about the theory of evolution in general and biological evolution in particular. They also have the responsibility to impart to their students an understanding of a scientific worldview and of the limits, possibilities, and dynamics of science as a way of knowing. Combining an understanding of the theory of evolution with an understanding of the nature of science is one way to think about the title of this compendium—*Evolution in Perspective*.

The articles in the compendium, selected to serve as resources for science teachers as they teach evolution, provide perspectives from scientific and educational leaders who understand evolution, the nature of science, and very importantly, the crucial role of science teaching. The articles equip science teachers with background information about evolution and the nature of science, the role of evidence, contemporary examples of evolution, and different perspectives that support the integrity of science and defend against assaults on the science curriculum. That said, however, teaching evolution is not without challenges.

Teaching evolution in perspective challenges educators to address continuing efforts to reduce the attention given to the topic of evolution in school science programs—or to eliminate it altogether. Science teachers face four critical challenges as they attempt to teach evolution. First, science teachers must introduce students to the scientific concepts of evolution together with, and as an integral part of, the nature of science. This calls for an examination of textbooks and teaching. Second, teachers must replace their defensive posture in the face of attacks from fundamentalist groups with constructive ap-

INTRODUCTION

proaches for teaching evolution. In the recent past, scientists and educators have expended considerable time and energy on the defensive—debating the ill-conceived, but often well-organized and articulate positions of creationists. Some states, schools, and teachers have acceded to the challenges of these groups by changing school science materials and programs in ways that undermine the integrity of science. To address this issue, science teachers must understand and teach the scientific theory of evolution. Third, closely related to the issue of teaching the theory of evolution is the need to understand and teach about inquiry and the nature of science. Finally, science teachers require the support of the broader scientific community if they are to meet these challenges.

TEXTBOOKS AND TEACHING

We would do well to begin with an examination of science textbooks. For the most part, our textbooks read like a collection of facts—small revealed truths that miraculously appear on pages without explanation. Students have no opportunity to learn how these scientific facts came to be or why they are different from any other way of explaining the natural world. At best, scientific processes are presented as lists of words such as "observation," "hypothesis," "inference," and so on. At worst, students memorize the scientific method as a five- or six-step procedure. Is it any wonder that students and adults don't understand inquiry and the nature of science? These are some general concerns about science textbooks. Specifically, how do textbooks address evolution and the nature of science?

Some good news can be reported about the topic of evolution in science textbooks. Although there remains a lingering impact of the 1925 Scopes trial on high school biology textbooks, mainstream biology textbooks include evolution. I am very pleased that history records the fact that the three textbooks produced in the early 1960s by the organization I direct, the Biological Sciences Curriculum Study (BSCS), reintroduced evolution as the unifying theme of biology. From 1925 until 1963 there had been a steady decline in the coverage of evolution in high school textbooks. BSCS put evolution back in the high school textbooks (Moore, 2001; Rudolph, 2002). BSCS continues this tradition with a new program, *BSCS Biology: A Human Approach* (BSCS, 2003), in which the first major unit is "Evolution: Patterns and Products of Change in Living Systems." Concerning textbooks, I think it reasonable to say that developers and publishers have responded to the need to include evolution as a topic. How it is presented remains an issue, and attention to inquiry and the nature of science continues to receive only marginal emphasis.

What about science teaching? Unfortunately, many science teachers avoid teaching the theory of evolution and using the inquiry approach, and the nature of science has little importance in today's science curricula. Teaching science in a manner that requires students to memorize facts and technical vocabulary does not help them understand major conceptual ideas such as evolution (Bransford, Brown, and Cocking, 2000; Bybee, 2002). To the degree some teach science in this manner, we will have to confront students' lack of understanding of science in general and evolution in particular. Many students never really learn that science is a way of explaining the natural world and a discipline characterized by careful observation and experimentation in which explanations are tentative, are open to skeptical review, and can change based on evidence. Moreover, they never experience the excitement and wonder of science—another outcome of an inadequate science education.

It is no mystery then that citizens confronting declarations and questions such as "Evolution is never observed," "Where are the missing links?" "Science cannot explain...," and "Evolution is only a theory" are confused about the critical distinction between science and religion and, subsequently, about support for science in the school curriculum. Take the last example: "Evolution is only a theory." In everyday discussions of nonscientific issues, individuals often say, "I have a theory about that." At best they mean "I have a personal explanation," and often they mean "I have an opinion." Seldom do they mean "I have evidence supporting my position." So, when confronted with the assertion that "evolution is only a theory," and the emphasis is on "only," the average person assumes that creationists' ideas have equal *scientific* standing with the theory of descent with modification. This issue is further confused because most individuals do not understand the relationship of scientific facts to theories. So they assume that facts are true and theories are only educated guesses. They simply do not understand the nature of science because they have not had an adequate science education.

TEACHING THE SCIENTIFIC THEORY OF EVOLUTION

The cultural contributions of science include a set of major ideas about how the natural world works. The particulate model of matter, the germ theory of infectious disease, and the gene model of inheritance are all examples of science's intellectual contributions to culture. The theory of biological evolution is among the most powerful and significant scientific contributions of the nineteenth and twentieth centuries (Mayr, 2000). This said, my earlier discussions of textbooks and teaching points out a fundamental educational problem. When teachers focus on numerous technical terms, students lose sight of the major idea. Metaphorically, students are examining seedlings and they fail to see the trees, much less the forest. With the focus we have in most current textbooks, the low levels of scientific literacy should not at all surprise us.

Science teachers can, for example, teach the theory of evolution using the approach of historical case study. This narrative form, or the explanatory story, provides students with a unique form of science education, and it affords science teachers the opportunity to integrate the nature of science and evolutional theory. Explanations related to the theme of evolution have a long and very rich history, and they center on interesting, often compelling, questions such as "Why are there so many different kinds of plants and animals?" and "How can the similarities among organisms be explained?" If educators built a portion of the science curriculum on the historical development of explanations to such questions, they would help to make the theory of evolution coherent and meaningful for students, equipping them to counter the continuous assaults by creationists.

The narrative approach can also help students understand the historical and scientific development of the theory of evolution. This approach, which emphasizes interrelated facts, propositions, inferences, and the human struggle to develop the strong explanations that form a theory such as evolution, provides an alternative to the memorization of details and single propositions. It also illustrates how theories are formulated and developed, thus helping students understand one of the most important activities in the growth of scientific knowledge (Lewis, 1988).

From a science teaching point of view, presentation of the formulation and development of the theory of evolution provides a logical structure for established knowledge, reveals logical relationships among the facts and inferences of a theory, and demonstrates the role that human imagination plays in the processes of science. Teachers can introduce students to a set of postulates and then continually pose questions about evidence supporting the postulates, the development of explanation, the ability to predict, and the lines of reasoning that connect the evidence, explanations, and predictions (Lewis, 1987).

In *One Long Argument*, Ernst Mayr (1991) provides an example of the relationship between em-

Darwin's Theory of Evolution by Natural Selection

Fact 1: All species have such great potential to produce large numbers of offspring that their population size would increase exponentially if all individuals that are born survived and reproduced successfully.

Fact 2: Except for seasonal fluctuations, most populations are normally stable in size.

Fact 3: Natural resources are limited, and in a stable environment they remain relatively constant.

  Inference 1: Because more individuals are produced than the available resources can support, and the population size remains stable, there must be a fierce struggle for existence among the individuals of a population. This results in the survival of only a part, often a very small part, of the offspring of each generation.

Fact 4: No two individuals in a population of organisms are exactly the same; rather, each population displays enormous variation in characteristics.

Fact 5: Much of this variation can be inherited.

  Inference 2: Survival in the struggle for existence is not random but depends in part on the characteristics that the surviving individuals inherited. This unequal survival is a process of natural selection that favors individuals with characteristics that best fit them in their environment.

  Inference 3: Through many generations, this process of natural selection will lead to a continuing, gradual change in populations—that is, to evolution and the production of new species.

Based on a description by Ernst Mayr. 1991. *One Long Argument*. Cambridge, MA: Harvard University Press.

pirical observations (facts) and the inferences that form a vital aspect for the scientific endeavor. Figure 1 presents an example of the logic, reasoning, and imagination that go into formulating connections between evidence and scientific explanations.

The recommendations expressed in this section are not new (Lewis, 1986). Further, there are excellent resources, for example, John A. Moore's *Science as a Way of Knowing* (1993), Ernst Mayr's *One Long Argument* (1991) and *This Is Biology* (1997), and I. B. Cohen's *Revolution in Science* (1985), that provide background and knowledge for the science teacher. In addition, science teachers will find *Teaching about Evolution and the Nature of Science* (National Academy of Sciences, 1998) a valuable resource.

TEACHING INQUIRY AND THE NATURE OF SCIENCE

When asked for a definition of science, most science teachers express the complementary ideas that science is both a body of knowledge and a process. In the course of such discussions few disagree with an assertion made by the late Carl Sagan (1993) that "Science is a way of thinking much more than it is a body of knowledge." Yet, as pointed out in the prior section, science textbooks and teaching give significantly more emphasis to science as a body of knowledge than to science as a way of thinking. The emergence of modern science in the late sixteenth and early seventeenth centuries was primarily due to the acceptance of new ways of thinking and explaining the natural world.

How does one characterize the basic elements of a scientific way of thinking? Briefly, a scientific explanation of nature must be based on empirical evidence from observations and experiments. Proposed explanations about how the world works must be tested against empirical evidence from nature. The scientific way of thinking stands in contrast to other ways of explaining nature—for example, religious doctrine or the acceptance of statements by arbitrary authority. After the scientific revolution such approaches to explaining the natural world were no longer satisfactory. Explanations had to be subject to confirmation by empirical evidence. For example, Galileo's observations of heavenly bodies confirmed Copernicus's heliocentric explanations of planetary motion. Since the emergence of modern science, our understanding of the natural world has been based on current explanations and on the interaction of human reasoning, imagination, and the empirical evidence of nature itself. One could reasonably argue that the scientific way of knowing also is among the great intellectual achievements. To emphasize points made earlier, students leave our schools without a clear understanding of the nature of scientific knowledge and the ways by which scientists claim to know about nature. Students may learn that science proceeds through a prescribed five-step method or through various laboratory procedures, but they do not learn about the processes of science. Thus, they lack a fundamental understanding of science. Teaching about inquiry and the nature of science could serve as a major countervailing force against those who propose that dogmatic, nonscientific explanations be included in the science curriculum.

SUPPORT FOR SCIENCE TEACHERS

Science teachers confront many challenges. In this introduction I have named several directly related to teaching the theory of evolution in perspective. Fortunately, however, support for teaching about evolution is available in a variety of forms.

National policy documents that address the content of school science programs clearly and directly promote the teaching of evolution. One document in particular provides both guidance for the content of school science programs and support for those who encounter challenges to scientific integrity and content of the science curriculum—the *National Science Education Stan-*

dards (NRC, 1996). Here teachers will find significant support for the inclusion of evolution, by name, in school science programs. In addition, science teachers will find consistency between statements in this document and two publications produced by the American Association for the Advancement of Science (AAAS)—*Science for All Americans* (1989) and *Benchmarks for Science Literacy* (1993). *Science for All Americans* includes clear and succinct discussions of the evolution of life, the nature of science, and the diversity of life. *Benchmarks* contains similar sections and provides brief statements of what individuals should know about the respective categories. *Benchmarks* and *Standards* have many similarities, and overlap significantly in some science content areas (AAAS, 1996). Teachers of science can use the consistency displayed by these two major scientific organizations, the National Research Council and the American Association for the Advancement of Science, as support for teaching evolution and the history and nature of science.

Depending on the state, science teachers have varied support from state standards. A report released in 2000 described and gave grades on states' treatment of evolution in science standards (Lerner, 2000). About half (24) of the states received A or B. Particularly noteworthy were California, Connecticut, Indiana, New Jersey, North Carolina, Rhode Island, South Carolina, Delaware, Hawaii, and Pennsylvania. Unfortunately, 13 states received F or F-. So, depending on the state, science teachers may or may not have the support they need. Personally, I think this is a national embarrassment and a dereliction of responsibility by states receiving any grade less than an A.

Important professional organizations for science teachers, specifically the National Science Teachers Association (NSTA), the National Association of Biology Teachers (NABT), and the National Earth Science Teachers Association (NESTA) have position statements that support the teaching of evolution. These position statements are complemented by those of scientific groups such as The National Academies, American Institute of Biological Sciences, American Geological Institute, American Association for the Advancement of Science, and other discipline specific groups.

I would be remiss if I did not mention the National Center for Science Education and the tireless work of Eugenie Scott and her colleagues. The various initiatives associated with creationism and intelligent design would have a much greater influence if it were not for Dr. Scott. Her support for science teachers is truly outstanding. Other individuals also have provided support— for example, Randy Moore, editor of *The American Biology Teacher*; Ken Miller, Brown University; Don Kennedy, Editor of the AAAS journal *Science*; and Bruce Alberts, president of the National Academy of Sciences, to mention a few.

CONCLUSION

In closing, I turn to this NSTA compendium. For some time NSTA has included articles on evolution in *The Science Teacher*, its journal for high school teachers. Publishing such articles, in my view, sends a clear and direct message of support to science teachers. I was honored when asked to recommend and organize articles dealing with evolution from past issues of *The Science Teacher*. Upon review I realized that the articles represented perspectives from scientists, science educators, and science teachers. This compendium represents some of the best contemporary scientific and educational thinking and presents evolution in perspectives that will be a resource and support for all science teachers.

REFERENCES

American Association for the Advancement of Science (AAAS). 1989. *Science for All Americans: A Project 2061 Report on Goals in Science, Mathematics, and Technology*. Washington, DC: Author.

American Association for the Advancement of Science (AAAS). 1993. *Benchmarks for Science Literacy*. Washington, DC: Author.

American Association for the Advancement of Science (AAAS). 1996. Creationists evolve new strategy. *Science News and Views* 273 (July 26): 420–422.

Biological Sciences Curriculum Study (BSCS). 2003. *BSCS Biology: A Human Approach*. Dubuque, IA: Kendall/Hunt Publishing Company.

Bransford, J.D., Brown, A.L., and Cocking, R.R. (Eds.). 2000. *How People Learn: Brain, Mind, Experience, and School*. Washington, DC: National Academy Press.

Bybee, R.W. (Ed.). 2002. *Learning Science and the Science of Learning*. Science Educator's Essay Collection. Arlington, VA: NSTA Press.

Cohen, I.B. 1985. *Revolution in Science*. Cambridge, MA: Harvard University Press.

Lerner, L.S. 2000. *Good Science, Bad Science: Teaching Evolution in the States*. Washington, DC: The Thomas B. Fordham Foundation.

Lewis, R.W. 1986. Teaching the theories of evolution. *The American Biology Teacher* 48(6): 334–347.

Lewis, R.W. 1987. Theories, concepts, mapping, and teaching. *The University Bookman: A Quarterly Review* 27(4): 4–11.

Lewis, R.W. 1988. Biology: A hypothetico-deductive science. *The American Biology Teacher* 50(6): 362–366.

Mayr, E. 1991. *One Long Argument*. Cambridge, MA: Harvard University Press.

Mayr, E. 1997. *This Is Biology*. Cambridge, MA: Harvard University Press.

Mayr, E. 2000. Darwin's influence on modern thought. *Scientific American* (July): 79–83.

Moore, J.A. 1993. *Science As A Way of Knowing*. Cambridge, MA: Harvard University Press.

Moore, R. 2001. The lingering impact of the Scopes trial on high school biology textbooks. *BioScience* 51(9): 790–796.

National Academy of Sciences. 1998. *Teaching About Evolution and the Nature of Science*. Washington, DC: National Academy Press.

National Research Council (NRC). 1996. *National Science Education Standards*. Washington, DC: National Academy Press.

Rudolph, J. L. 2002. *Scientists in the Classroom: The Cold War Reconstruction of American Science Education*. New York: Palgrave.

Sagan, C. 1993. *Broca's Brain: Reflections on the Domain of Science*. New York: Ballantine Books/Mass Market Paperback.

NSTA POSITION STATEMENT

The Teaching of Evolution

INTRODUCTION
The National Science Teachers Association (NSTA) strongly supports the position that evolution is a major unifying concept in science and should be included in the K–12 science education frameworks and curricula. Furthermore, if evolution is not taught, students will not achieve the level of scientific literacy they need. This position is consistent with that of the National Academies, the American Association for the Advancement of Science (AAAS), and many other scientific and educational organizations.

NSTA also recognizes that evolution has not been emphasized in science curricula in a manner commensurate with its importance because of official policies, intimidation of science teachers, the general public's misunderstanding of evolutionary theory, and a century of controversy. In addition, teachers are being pressured to introduce creationism, "creation science," and other nonscientific views, which are intended to weaken or eliminate the teaching of evolution.

DECLARATIONS
Within this context, NSTA recommends that
- Science curricula, state science standards, and teachers should emphasize evolution in a manner commensurate with its importance as a unifying concept in science and its overall explanatory power.
- Science teachers should not advocate any religious interpretations of nature and should be nonjudgmental about the personal beliefs of students.
- Policy makers and administrators should not mandate policies requiring the teaching of "creation science" or related concepts, such as so-called "intelligent design," "abrupt appearance," and "arguments against evolution." Administrators also should support teachers against pressure to promote nonscientific views or to diminish or eliminate the study of evolution.

NSTA POSITION STATEMENT

- Administrators and school boards should provide support to teachers as they review, adopt, and implement curricula that emphasize evolution. This should include professional development to assist teachers in teaching evolution in a comprehensive and professional manner.
- Parental and community involvement in establishing the goals of science education and the curriculum development process should be encouraged and nurtured in our democratic society. However, the professional responsibility of science teachers and curriculum specialists to provide students with quality science education should not be compromised by censorship, pseudoscience, inconsistencies, faulty scholarship, or unconstitutional mandates.
- Science textbooks shall emphasize evolution as a unifying concept. Publishers should not be required or volunteer to include disclaimers in textbooks that distort or misrepresent the methodology of science and the current body of knowledge concerning the nature and study of evolution.

—Adopted by the
NSTA Board of Directors
July 2003

NSTA offers the following background information:

THE NATURE OF SCIENCE AND SCIENTIFIC THEORIES

Science is a method of explaining the natural world. It assumes that anything that can be observed or measured is amenable to scientific investigation. Science also assumes that the universe operates according to regularities that can be discovered and understood through scientific investigations. The testing of various explanations of natural phenomena for their consistency with empirical data is an essential part of the methodology of science. Explanations that are not consistent with empirical evidence or cannot be tested empirically are not a part of science. As a result, explanations of natural phenomena that are not based on evidence but on myths, personal beliefs, religious values, and superstitions are not scientific. Furthermore, because science is limited to explaining natural phenomena through the use of empirical evidence, it cannot provide religious or ultimate explanations.

The most important scientific explanations are called "theories." In ordinary speech, "theory" is often used to mean "guess" or "hunch," whereas in scientific terminology, a theory is a set of universal statements that explain some aspect of the natural world. Theories are powerful tools. Scientists seek to develop theories that

- are firmly grounded in and based upon evidence;
- are logically consistent with other well-established principles;
- explain more than rival theories; and
- have the potential to lead to new knowledge.

The body of scientific knowledge changes as new observations and discoveries are made. Theories and other explanations change. New theories emerge, and other theories are modified or discarded. Throughout this process, theories are formulated and tested on the basis of evidence, internal consistency, and their explanatory power.

EVOLUTION AS A UNIFYING CONCEPT

Evolution in the broadest sense can be defined as the idea that the universe has a history: that change through time has taken place. If we look today at the galaxies, stars, the planet Earth, and the life on planet Earth, we see that things today are different from what they were in the past: galaxies, stars, planets, and life forms have evolved. Biological evolution refers to the scientific theory that living things share ancestors from which they have diverged; it is called "descent with modification."

There is abundant and consistent evidence from astronomy, physics, biochemistry, geochronology, geology, biology, anthropology, and other sciences that evolution has taken place.

As such, evolution is a unifying concept for science. The *National Science Education Standards* recognizes that conceptual schemes such as evolution "unify science disciplines and provide students with powerful ideas to help them understand the natural world" (NRC, 1996, 104) and recommends evolution as one such scheme. In addition, *Benchmarks for Science Literacy* from AAAS's Project 2061, as well as other national calls for science reform, all name evolution as a unifying concept because of its importance across the disciplines of science. Scientific disciplines with a historical component, such as astronomy, geology, biology, and anthropology, cannot be taught with integrity if evolution is not emphasized.

There is no longer a debate among scientists about whether evolution has taken place. There *is* considerable debate about *how* evolution has taken place: What are the processes and mechanisms producing change, and what has happened specifically during the history of the universe? Scientists often disagree about their explanations. In any science, disagreements are subject to rules of evaluation. Scientific conclusions are tested by experiment and observation, and evolution, as with any aspect of theoretical science, is continually open to and subject to experimental and observational testing.

The importance of evolution is summarized as follows in the National Academy of Sciences publication *Teaching about Evolution and the Nature of Science*: "Few other ideas in science have had such a far-reaching impact on our thinking about ourselves and how we relate to the world" (NAS, 1998, 21).

CREATIONISM AND OTHER NONSCIENTIFIC VIEWS

The *National Science Education Standards* notes that "[e]xplanations of how the natural world changes based on myths, personal beliefs, religious values, mystical inspiration, superstition, or authority may be personally useful and socially relevant, but they are not scientific" (NRC, 1996, 201). Because science limits itself to natural explanations and not religious or ultimate ones, science teachers should neither advocate any religious interpretation of nature nor assert that religious interpretations of nature are not possible.

The word "creationism" has many meanings. In its broadest meaning, creationism is the idea that the universe is the consequence of something transcendent. Thus to Christians, Jews, and Muslims, God created; to the Navajo, the Hero Twins created; for Hindu Shaivites, the universe comes to exist as Shiva dances. In a narrower sense, "creationism" has come to mean "special creation": the doctrine that the universe and all that is in it were created by God in essentially its present form, at one time. The most common variety of special creationism asserts that

- the Earth is very young,
- life was created by God,
- life appeared suddenly,
- kinds of organisms have not changed since the creation, and
- different life forms were designed to function in particular settings.

This version of special creation is derived from a literal interpretation of Biblical Genesis. It is a specific, sectarian religious belief that is not held by all religious people. Many Christians and Jews believe that God created through the process of evolution. Pope John Paul II, for example, issued a statement in 1996 that reiterated the Catholic position that God created and affirmed that the evidence for evolution from many scientific fields is very strong.

"Creation science" is a religious effort to support special creationism through methods of science. Teachers are often pressured to include it or other related nonscientific views such as "abrupt appearance theory," "initial complexity theory," "arguments against evolution," or "intelligent design theory" when they teach evolution. Scientific creationist claims have been discredited by the available scientific evidence. They have no empirical power to explain the natural world and its diverse phenomena. Instead, creationists seek out supposed anomalies among many existing theories and accepted facts. Furthermore, "creation science" claims do not lead to new discoveries of scientific knowledge.

LEGAL ISSUES

Several judicial decisions have ruled on issues associated with the teaching of evolution and the imposition of mandates that "creation science" be taught when evolution is taught. The First Amendment of the Constitution requires that public institutions such as schools be religiously neutral; because "creation science" asserts a specific, sectarian religious view, it cannot be advocated in the public schools.

When Arkansas passed a law requiring "equal time" for "creation science" and evolution, the law was challenged in Federal District Court. Opponents of the bill included the religious leaders of the United Methodist, Episcopalian, Roman Catholic, African Methodist Episcopal, Presbyterian, and Southern Baptist churches, along with several educational organizations. After a full trial, the judge ruled that "creation science" did not qualify as a scientific theory (*McLean v. Arkansas Board of Education*, 529 F. Supp. 1255 [ED Ark. 1982]).

Louisiana's equal time law was challenged in court, and eventually reached the Supreme Court. In *Edwards v. Aguillard* [482 U.S. 578 (1987)], the court determined that "creation science" was inherently a religious idea and to mandate or advocate it in the public schools would be unconstitutional. Other court decisions have upheld the right of a district to require that a teacher teach evolution and not teach "creation science" (*Webster v. New Lennox School District* #122, 917 F.2d 1003 [7th Cir. 1990]; *Peloza v. Capistrano Unified School District*, 37 F.3d 517 [9th Cir. 1994]).

Some legislators and policy makers continue attempts to distort the teaching of evolution through mandates that would require teachers to teach evolution as "only a theory" or that require a textbook or lesson on evolution to be preceded by a disclaimer. Regardless of the legal status of these mandates, they are bad educational policy. Such policies have the effect of intimidating teachers, which may result in the de-emphasis or omission of evolution. As a consequence, the public will only be further confused about the nature of scientific theories. Furthermore, if students learn less about evolution, science literacy itself will suffer.

REFERENCES

American Association for the Advancement of Science (AAAS), Project 2061. (1993). *Benchmarks for Science Literacy*. New York: Oxford University Press.

National Academy of Sciences . 1998. *Teaching About Evolution and the Nature of Science*. Washington, DC: Steering Committee on Science and Creationism, National Academy Press.

National Research Council. 1996. *National Science Education Standards*. Washington, DC: National Academy Press.

ADDITIONAL RESOURCES

Laudan, L. 1996. *Beyond Positivism and Relativism: Theory, Method, and Evidence*. Boulder, CO: Westview Press.

National Academy of Sciences. (1999). *Science and Creationism: A View from the National Academy of Sciences*, Second Edition. Washington, DC: National Academy Press.

Ruse, M. 1996. *But Is It Science: The Philosophical Question in the Creation/Evolution controversy*. Amherst, NY: Prometheus.

Skehan, J. W., and Nelson, C.E. 1993. *The Creation Controversy and the Science Classroom*. Arlington, VA: National Science Teachers Association.

ARGUING FOR EVOLUTION

FRANCISCO J. AYALA

n August 11, 1999, the Kansas State Board of Education voted six to four to remove references to cosmology and evolution from the state's education standards and assessments. The board's decision does grave disservice to the students and teachers of the state of Kansas as well as to science and religion everywhere. Students need to study the empirical evidence and concepts central to scientific knowledge to become informed and responsible citizens and to acquire suitable job skills and professional training. The board's decision places Kansas's students at a competitive disadvantage in their education and job qualification and impairs the recruitment of capable and inspired teachers, who will abhor being inhibited from teaching their best knowledge.

EVOLUTION AND SCIENCE

Opponents to teaching the theory of evolution declare that it is only a theory and not a fact; and that science relies on observation, replication, and experimentation, but nobody has seen the origin of the universe or the evolution of species, nor have these events been replicated in the laboratory or by experiment.

When scientists talk about the "theory" of evolution, they use the word differently than people do in ordinary speech. In everyday speech, a theory is considered to be an imperfect fact, as in "I have a theory as to what caused the explosion of TWA flight 800." In science, however, a theory is based on a body of knowledge.

According to the theory of evolution, organisms are related by common descent. There is a multiplicity of species because organisms change from gen-

THE SCIENTIFIC PERSPECTIVE

eration to generation, and different lineages change in different ways. Species that share a recent ancestor are therefore more similar than those with remote ancestors. Thus, humans and chimpanzees are, in configuration and genetic makeup, more similar to each other than they are to baboons or to elephants.

Scientists agree that the evolutionary origin of animals and plants is a scientific conclusion beyond reasonable doubt. They place it beside such established concepts as the roundness of the Earth, its rotation around the Sun, and the molecular composition of matter. That evolution has occurred, in other words, is a fact.

How is this factual claim compatible with the accepted view that science relies on observation, replication, and experimentation, even though nobody has observed the evolution of species, much less replicated it by experiment? What scientists observe are not the concepts or general conclusions of theories, but their consequences. Copernicus's heliocentric theory affirms that the Earth rotates around the Sun. Nobody has observed this phenomenon, but we accept it because of numerous confirmations of its predicted consequences. We accept that matter is made of atoms, even though nobody has seen them, because of corroborating observations and experiments in physics and chemistry. The same is true of the theory of evolution. For example, the claim that humans and chimpanzees are more closely related to each other than they are to baboons leads to the prediction that the DNA is more similar between humans and chimps than between chimps and baboons. To test this prediction, scientists select a particular gene, examine its DNA structure in each species, and thus corroborate the inference.

> Holding strong religious beliefs does not preclude intelligent scientific thinking.

Experiments of this kind are replicated in a variety of ways to gain further confidence in the conclusion. And so it is for myriad predictions and inferences among all sorts of organisms.

Not every part of the theory of evolution is equally certain. Many aspects remain subject to research, discussion, and discovery. But uncertainty about these aspects does not cast doubt on the fact of evolution. Similarly, we do not know all the details about the configuration of the Rocky Mountains and how they came about, but this is no reason to doubt that the Rockies exist.

The theory of evolution needs to be taught in schools because nothing in biology makes sense without it. Modern biology has broken the genetic code, developed highly productive crops, and provided knowledge for improved health care. Students need to be properly trained in biology to improve their education, increase their chances for gainful employment, and enjoy a meaningful life in a technological world.

RELIGION AND SCIENCE

Does the teaching of evolution pose a threat to Christianity or to other religions? This question can be answered in two parts. I would first address those who profess a materialistic philosophy and seek to ground it in the theory of evolution and other scientific claims. They point out the great success of science in explaining the workings of the universe and claim there is no room for other kinds of explanations—no room for values, morality, or religion. We may grant these persons their right to think as they wish, but they have no warrant whatsoever to ground this materialistic philosophy in the achievements of science. Science seeks material explanations for material processes, but it has nothing definitive to say about realities beyond its scope. Science is a way of acquiring knowledge about ourselves and the world around us, but it is not the only way. We acquire knowledge in many other ways, such as through literature, the arts, philosophical reflection, and religious experience. Scientific knowledge may enrich aesthetic and moral perceptions, but these subjects transcend science's realm.

Scientific knowledge cannot contradict religious beliefs, because science has nothing to say for or against religious realities or religious values. Many religious authorities have made this point. Catholic, Lutheran, and other Christian bishops have joined Jewish and other religious leaders in denying that the theory of evolution contradicts or threatens their religious beliefs.

There are, however, believers who see the theory of evolution and scientific cosmology as contrary to the creation narrative of the Book of Genesis. We may grant these believers their right to think this, just as at the other extreme of the spectrum we grant materialists their right to deny spiritual or religious values. But, as the counterpoint to what I said above, I will aver that Genesis is a book of religious revelations, not a textbook of astronomy or biology. Pope John Paul II has made the point: "The Bible speaks to us of the origins of the universe and its makeup, not in order to provide us with a scientific treatise, but in order to state the correct relationship of man with God and the universe. Sacred scripture wishes simply to declare that the world was created by God, and in order to teach this truth, it expresses itself in the terms of the cosmology in use at the time of the writer. The sacred book likewise wishes to tell men that the world was ... created for the service of man and the glory of God. Any other teaching about the origin and makeup of the universe is alien to the intentions of the Bible, which does not wish to teach how heaven was made but how one goes to heaven." St. Augustine had made the point many centuries earlier: "In the matter of the shape of heaven the sacred writers did not want to teach men facts that would be of no avail for their salvation." (See note, p. 4.)

The point made by St. Augustine and the pope is that it is a blunder to mistake the Bible for an elementary textbook of astronomy, geology, and biology. Instead, it is possible to believe that God created the world while also accepting that the planets, mountains, plants, and animals came about, after the initial creation, by natural processes. I can believe that I am God's creature without denying that I developed from a single cell in my mother's womb by natural processes. This is the second part of my answer to the purported opposition between scientific conclusions and religious beliefs. They do not stand in contradiction; they concern different sorts of issues, belong to different realms of knowledge.

INTELLIGENT DESIGN?

There is one more point I wish to make in response to those who defend the special creation of the species on the grounds of their design, which they see as necessarily a product of divine intelligence. The point is that not only can natural selection account for the "design" of organisms but also it amounts to blasphemy to attribute it to God's special action.

Consider the human jaw. We have too many teeth for the jaw's size, so that wisdom teeth need to be removed, and orthodontists make a decent living straightening the other teeth. Would we want to blame God for such defective design? A human engineer could have done better. Evolution gives a good account of this imperfection. Brain size increased over time in our ancestors, and the remodeling of the skull to fit the larger brain entailed a reduction of the jaw. Evolution responds to the organism's needs through natural selection, not by optimal design, but by "tinkering" as it were, by slowly modifying existing structures. Consider now the birth canal of women, much too narrow for easy passage of the infant's head, so that thousands upon thousands of babies die during delivery. Surely we do not want to blame God for this defective design or for the children's deaths. Science makes it understandable, a consequence of the evolutionary enlargement of our brain. Females of other animals do not experience this difficulty.

One more example: why are our arms and our legs, which are used for such different functions, made of the same materials, the same bones, muscles, and nerves, all arranged in the same overall pattern? Evolution makes sense of the anomaly. Our remote ancestors' forelimbs were legs. After our ancestors became bipedal and started using their forelimbs for functions other than walking, these became gradually modified, but retained their original composition and arrangement. Engineers start with raw materials and a design suited for a particular purpose; evolution can only modify what is already there. An engineer who designed cars and airplanes, or wings and wheels, using the same materials arranged in a similar pattern, would surely be fired. The defective design of organisms could be attributed to the gods of the ancient Greeks, Romans, and Egyptians, who fought with one another, made blunders, and were clumsy in their endeavors. But in my view, it is not compatible with a special action by the omniscient and omnipotent God of Judaism, Christianity, and Islam.

There is no need for warfare between science and religion. It is unfortunate that some would deprive students of a proper scientific education on religious grounds, as it is unfortunate that some seek in science arguments to negate the legitimacy of religious beliefs.

BIOGRAPHICAL NOTE

John Paul II's quotation is from his address to the Pontifical Academy of Sciences on October 3, 1981. In his address to the Pontifical Academy of Sciences on October 22, 1996, he again deplored interpreting the Bible's teachings as scientific rather than religious, and said: "[N]ew knowledge has led us to realize that the theory of evolution is no longer a mere hypothesis. It is indeed remarkable that this theory has been progressively accepted by researchers, following a series of discoveries in various fields of knowledge. The convergence, neither sought nor fabricated, of the results of work that was conducted independently is in itself a significant argument in favor of this theory" (*L'Osservatore Romano*, 23 October 1996). St. Augustine's quotation is from *The Literal Meaning of Genesis,* book 2, chapter 9. In book 3, chapter 14, he makes the remarkable statement that many animal species were not present from the beginning, but appeared later "each according to its kind and with its special properties," as a result of a natural power "present from the beginning in all living bodies." One can surmise that Augustine would have found no conflict between the theory of evolution and the teachings of Genesis, which are the subject of his commentary.

EVIDENCE FOR EVOLUTION

G. BRENT DALRYMPLE

ver the past several decades a small group of evangelical Christians, best described as young-Earth creationists, have made a series of outrageous and untrue assertions about science and the natural world. In particular, they dispute the history of the Earth, the universe, and Earth's biota as scientists currently understand those subjects, and they claim, among other things, that Earth is extremely young. They say their beliefs about the origin and history of the natural world, which they bill as "scientific creationism," are every bit as scientific as those of real science, which they commonly misname "evolution science." Using arguments of equality and fairness, young-Earth creationists wrongly insist that their version of natural history should be taught in science classes.

"Scientific creationism," an oxymoron, is religion pure and simple, a fact clearly recognized by federal court rulings in both Arkansas (*McLean v. Arkansas Board of Education*, 1982) and Louisiana (*Aguillard v. Treen*, 1985). Federal district court judges ruled in both cases that "creation science" is not science but religion, and struck down as unconstitutional the "equal time for creationism" laws of those states. The U.S. Supreme Court subsequently affirmed the Louisiana decision (*Edwards v. Aguillard*, 1987). The National Academy of Sciences (NAS) recently reaffirmed its 1984 position that "creation science" is not science (Steering Committee on Science and Creationism, 1999), and many scientific and religious organizations have taken similar positions (Matsumura,

1995). One only need glance at the literature of "scientific creationism" to see why the courts ruled, and NAS and others concluded, as they did.

The plain truth is that the young-Earth creationists have no valid data or calculations to support their vague claims that Earth is only 6000 to 12 000 years old, or thereabouts. Their young ages for Earth come solely from their belief in a literal six-day creation and an interpretation of the Biblical genealogies, which they freely admit in some of their less-guarded publications (Morris, 1994). In their attempts to refute contrary scientific data, the best they can do is to question the validity of scientific methods using irrelevant data or faulty logic. In contrast, there is a preponderance of scientific evidence that the Earth and solar system are approximately 4.54 ± 0.02 billion years old and that the Milky Way Galaxy and the universe are even older. The creationists' pseudoscientific arguments for a young Earth have been dealt with in numerous publications, and I will not rehash them here. Instead, I will ignore the creationists' mischievous nonsense and briefly review the scientific evidence for the age of the Earth, solar system, Galaxy, and universe, all of which is covered in more detail in *The Age of the Earth* (Dalrymple, 1991).

DATING THE EARTH

Scientists have more than half a dozen good ways of measuring the ages of rocks, particularly rocks that have once been molten, like lava and granite. These measurement methods, collectively called radiometric dating, rely on the decay of long-lived radioactive isotopes (with half-lives greater than a billion years), which occur naturally in virtually all rocks, and the corresponding accumulation of the stable isotopes produced by the decays. These radioisotope "clocks" are individually known by the chemical symbols of the elements of their parent–daughter pairs (U–Pb, K–Ar, Rb–Sr, and so forth). Modern techniques are so good that the precision of such measurements is often better than 1 percent, and scientists have even developed clever ways to make the methods self-checking against the possibility of errors due to unrecognized geologic events, like reheating.

Dating Earth directly is difficult because the planet is constantly changing. Earth's crust is continually being added to, modified, and destroyed. As a result, rocks that record Earth's earliest history have not been found and may no longer exist. Nevertheless, the oldest rocks on Earth provide at least a minimum estimate of the planet's age. Rocks older than 3.5 billion years (Ga, or giga annum) have been found and dated on all continents, but the oldest are in Greenland and in Canada. The Greenland rocks have been dated, many times, by four independent radiometric dating methods (U–Pb, Pb–Pb, Rb–Sr, Sm–Nd) at 3.7 to 3.8 Ga. These are metamorphic rocks that were originally sediments and lava. The sedimentary rocks tell us that even older rocks, the source for the sediments, existed at one time. In Canada, near Great Slave Lake, there are rocks that are nearly 4.0 Ga, and in Australia tiny grains of the mineral zircon found in younger sedimentary rocks have U–Pb ages of 4.0 to 4.2 Ga. From the hundreds of age measurements that have been made on Earth's most ancient surviving rocks, we know that our planet is more than 4.0 Ga.

The Moon has older rocks than Earth. Because the Moon is a small planet, the internal heat required for crustal recycling is long gone, and it has been a dead planet for some time. More than a hundred of the lunar rocks returned by the Apollo and Luna missions have been dated, using primarily two dating methods (K–Ar and Rb–Sr), and most are more than 3 Ga. There are a few, however, that have ages of 4.4 to 4.5 Ga,

Topic: Age of the Earth/Universe
Go to: *www.scilinks.org*
Code: EIP01

How old are the Earth, the Galaxy, and the Universe?

which tells us that the Moon is at least that old.

The most primitive objects in the Solar System are meteorites, which are fragments of asteroids. Thousands of meteorites have been recovered on Earth and more than a hundred have been dated by six different radioactive dating methods; a number of meteorites have been dated by two or more methods, with consistent results. The results show that the major meteorite types were formed within a few tens of millions of years between 4.5 and 4.6 Ga.

Thus, there is clear and indisputable evidence that the planets of the inner solar system, and presumably the entire solar system, originated 4.5 to 4.6 Ga. The number that scientists think best represents the age for the Earth and solar system is 4.54 ± 0.02 Ga. This age is based on the time required for the lead isotopes in four very old lead ores on Earth to have evolved, through the decay of uranium, from the composition of lead at the time the solar system formed. This "primordial" composition of lead is recorded in the Canyon Diablo meteorite in a particular mineral (troilite, a form of iron sulfide) that contains no uranium, so its lead composition could not have changed since the meteorite formed. All of the objects that can be dated by radiometric techniques come from the inner solar system, so other methods have to be employed to measure the ages of the Galaxy and universe.

THE BIG BANG

In the late 1920s, astronomer Edwin Hubble observed that distant galaxies were moving away from the Milky Way Galaxy in every direction, and the more distant the galaxies, the faster they were receding. He concluded that this could only be true if the universe were expanding. Hubble realized that this distance–velocity relationship could be extrapolated backward to the time when all the galaxies were at a single point and that such a calculation would provide an estimate of the time that has passed since the Big Bang. Since Hubble's discovery, many astronomers have verified and refined his observations so that the expansion of the universe is no longer in doubt, although there are several uncertainties involved in calculating an age from the universe from the velocity–distance measurements.

Velocity measurements are relatively easy to make accurately, but distances in the universe are very hard to measure with similar accuracy. There also is some uncertainty about whether the expansion has been uniform over time, or has been accelerating or slowing down. Over the past few decades there have been many estimates made of the age of the universe using the Hubble expansion, with a range of 7 to 20 Ga. The best estimates using the most precise data gathered over the past few years all fall within the range 10 to 15 Ga.

STAR DATES

The age of the Milky Way Galaxy has also been estimated. Surrounding the center of the Galaxy are several hundred star clusters—each containing many thousands of stars—called *globular clusters* because of their quasi-spherical shape. Because of their low metal contents, globular cluster stars are known to be some of the oldest stars in the Galaxy. Stars of different mass consume their nuclear fuel and evolve at different rates. The most massive

Topic: Edwin Hubble
Go to: *www.scilinks.org*
Code: EIP02

stars last less than 10 million years, whereas ordinary stars like the Sun last billions of years. As these stars evolve, their brightness and color also change, so astronomers can measure the degree of evolution of stars of different mass in a globular cluster. This information can then be compared with theoretical models of star evolution to estimate the age of the globular cluster and, hence, the age of the Galaxy. There have been many studies of globular clusters and all have resulted in estimates in the range of 11 to 16 Ga.

Another method for estimating the age of the Galaxy is based on the abundances in the solar system of certain long-lived radioactive isotopes. The abundance of these isotopes is a function of their production rates in supernovas. These "explosions" not only manufacture but distribute elements widely for later incorporation into other stars. Estimates of the age of the Galaxy based on the isotope abundance range from 9 to 16 Ga.

Thus, a preponderance of scientific evidence indicates that the universe and the Milky Way Galaxy are approximately 10 to 15 Ga, and the age of the solar system and its planets, including Earth, is 4.54 Ga. There is no scientific evidence whatsoever for ages of only a few thousand years.

So what should be taught in the public schools—the scientific version of the history of the Earth and universe or the creationists' version? That depends on the purpose of the course. If, on the one hand, science education is the goal, then the evidence and conclusions of real science, not pseudoscience, should be taught. Creationism, which is not science, has no place in such a curriculum. On the other hand, creationism might be a valid topic in classes in comparative religion, the history of religion, or philosophy. Many creationists advocate teaching both in science classes, but that makes no sense. It would be a great disservice to students to teach one subject (the religion of creationism) disguised as another (science). It would not only be confusing, but would result in an inferior education, and no dedicated educator wants that.

REFERENCES

Aguillard v. Treen, 634 F. Supp. 436 (ED Louisiana. 1985).

Dalrymple, G.B. 1991. *The Age of the Earth*. Stanford, Calif.: Stanford University Press.

Edwards v. Aguillard, 482 U.S. 578 (1987).

Matsumura, M. 1995. *Voices for Evolution*. Berkeley, Calif.: National Center for Science Education.

McLean v. Arkansas Board of Education, 529 F. Supp. 1255 (ED Arkansas. 1982).

Morris, J.D. 1994. *The Young Earth*. Colorado Springs, Col.: Creation-Life Publishers.

Steering Committee on Science and Creationism. 1999. *Science and Creationism: A View from the National Academy of Sciences*. Washington, D.C.: National Academy Press.

Wakefield, J.R. 1988. The geology of Gentry's "tiny mystery." *Journal of Geologic Education* 36(3):161–175.

THE SCIENTIFIC PERSPECTIVE

SECTION I

THOUGHT PATTERNS IN SCIENCE AND CREATIONISM

Understanding the vast difference between creationism and evolutionary theory

JOHN A. MOORE

The concept of evolution has been an important part of biology curricula, especially in high school courses, since the publication of the Biological Sciences Curriculum Study textbooks in the 1960s. Nevertheless, teaching the subject continues to be controversial because of the efforts of creationists to have "creation science" taught as well or to otherwise diminish the treatment of evolution. The influence of creationists seems to be increasing, as evidenced by the smoke rising over the plains of Kansas in the summer of 1999.

How can this be? The answer is twofold. First, teachers fail to provide very young students with an understanding of the way science works and especially of what counts for evidence. Second, we ignore the fact that for decades concerned parents and other citizens have expressed great interest in understanding the evidence for evolution and creationism—for an honest and dispassionate discussion of the issues so they can develop an informed opinion. So far, neither the scientific community nor schools have provided such an overview.

The conflict between scientists and creationists is fundamentally about systems of beliefs, especially about what constitutes acceptable evidence. Science

THE SCIENTIFIC PERSPECTIVE

is the study of the natural world, and statements made about it are based on data derived from observations and experiments. These statements must be related to the existing corpus of scientific knowledge and must be of such a nature that they can be confirmed, modified, or rejected by other scientists. They must never be accepted as final truth but only as representing the most accurate statement that can be made on the basis of available evidence.

In many instances, the level of certainty is such that some statements can be accepted as true beyond all reasonable doubt. For instance, we know that if we let go of an object on Earth, it will drop. We also know that it will not drop in a space vehicle. We explain these observations by saying that on Earth a force, gravity, attracts the objects; but in outer space the force of the Earth's gravity does not have an effect. Scientific statements such as this one are explanations of natural things and processes. They explain rather than dictate, in contrast with the laws of human society. Thus, laws of science are descriptive; the laws of society are proscriptive. Biologists tend not to use the word "law" but, instead, use the word "concept." For biologists, the concept of evolution is the explanation for the variety of species populating the Earth.

The belief system of creationists is fundamentally different. For them, the diversity of life is a consequence of a divine act in the not too distant past (maybe 4000 B.C.). A supernatural act of a god created all the kinds of living organisms that we see today and that lived in the past. Thus, creationists employ a different methodology than scientists do in explaining the diversity of life over time. Creationists begin with a conclusion, that a divine being created life, and then seek data to support that conclusion. The data consist of one of the accounts of creation, the so-called P version, which appears in the first two chapters of Genesis in the Judeo-Christian Bible. No other religious tradition has that specific story of a creation, and leaders of most of the major Jewish and Christian sects do not accept the Genesis account as scientifically valid. Since the last decades of the 1800s there has been a steady decline in the number of theologians and other biblical scholars who accept the Bible as a reliable source of scientific information. Instead, the Bible is regarded as a profound source of deep philosophical understanding of human life and a system of morals that can lead to a virtuous life.

> There can be no compromise between scientists and creationists because what is accepted as evidence by one group differs fundamentally from what is accepted by the other.

VALIDATING EVIDENCE

We have, therefore, two irreconcilable patterns of thought. Scientists must base their statements on data derived from natural things and processes. The goal is to arrive at a naturalistic way of explaining these natural things and processes. Thus, a huge number of facts relating to fossils of different ages, comparative anatomy, embryonic development, comparative genetics, comparative physiology, geographic distribution, the classification of species, and the experimental manipulation of populations under controlled conditions are found to "make sense" as being the consequences of evolution. No attempt to discredit the concept of evolution has proved successful. Thus, evolution is accepted as the most accurate statement that can be made based on the available data. The counter-explanations of Christian fundamentalists are rejected because a scientist cannot accept any explanations based on supernatural things and processes—the domain of science consists only of natural things and processes that can be studied and, hence, nothing can be said about the existence or non-existence of gods or their actions.

There can be no compromise between scientists and creationists because what is accepted as evidence by one group differs fundamentally from what is accepted by the other. For scientists, there is one science of evolutionary biology to which scientists of all faiths, or no religious faith at all, make their contributions. These evolutionary biologists constantly study new problems, test existing ideas in evolutionary biology, and explain new data in terms of current understanding in the field. Although there is a general consensus of the fundamental principles of evolutionary biology, there is also active debate among those working at the cutting edge of the field. On this boundary, between what is known and ignorance, new hypotheses are advanced, tested, disputed, rejected, or validated, and where possible, these hypotheses become part of the corpus of scientific knowledge.

In contrast, there is no way to validate the views about nature of one religious sect or another. Deeply religious persons accept the tenets of their faith on belief, not hard evidence. It is not even possible to provide a universally accepted definition of God. If humans were made in the image of God, God must have the structure and general nature of a human being. Others reject that as a primitive position and suggest that God is a divine intelligence and/or a force that is everywhere and in all things. One can accept the beliefs of any of the sub-sects of Hinduism, Buddhism, Islam, Judaism, or Christianity with little fear that any one will be proven false—or true for that matter. Many of the tragedies of human history, even today, are the result of sanguine conflicts between those professing different religious beliefs.

Thus, scientists use rational approaches with the essential property that all explanations must be based on natural phenomena and must be confirmable by any person who repeats the observations or experiments. Tests of relevant hypotheses, therefore, have the important function not only of increasing the explanatory power of the concept of evolution but also of eliminating errors. Science is a self-correcting process that is constantly testing its basic statements. For example, in the field of evolutionary biology there are instances of "discoveries" turning out to be incorrect. Creationists often cite the case of the Piltdown Man, reported to be parts of a fossil human skull found in Southern England, and a fossil tooth discovered in the Western United States, thought to be of a prehistoric ape called *Hesperopithecus*. Piltdown Man was found to be a fraud and *Hesperopithecus* a pig. Neither these, nor any other errors I know of, have been corrected by creationists—all of them have been corrected by scientists.

ACCEPTING EMPIRICAL EVIDENCE

Most readers accept the statements in science textbooks, general books on science, and the "just discovered" stories in the mass media. For most fields of knowledge, individuals tend to accept what authorities in the field say. This is usually true in the sciences but less so in those disciplines such as the humanities in which much is based on opinion. Few scientific explanations can be understood by everyone—even by other scientists in different fields. Physicists, for example, accept that the best scientific explanation for the origin of the universe, based on available data, is that there was a Big Bang about 12 billion years ago. Yet how many readers know the evidential basis for this statement or can evaluate the probability of it being correct?

Similarly, many people find the enormous age of the Earth fascinating. The estimates are based on radiometric methods for determining the time that sedimentary rocks were formed and therefore the age of any fossils found within that layer of rocks. Such data are absolutely essential for determining the changes in life over time. Radiometric methods have been confirmed and, most importantly, different radiometric methods give similar dates for the same rocks. The oldest known rocks are about 3.8 billion years old. Most creationists say they simply do not believe the data. They cannot, of course, when their faith requires them to believe that the world is less than 10 000 million years old.

Thus, creationists deny the empirical evidence that the scientific community accepts as supporting the concept of evolution beyond all reasonable doubt. Instead, creationists accept the supernatural explanation of divine creation for which there is no scientific evidence whatsoever. The lack of evidence for their own beliefs dictates that they try to discredit the concept of evolution in the mistaken and illogical argument that, if evolution is wrong, creationism must be right.

It has been suggested that the problem for many creationists is not the concept of evolutionary change so much as the suggestion that humans and apes are descended from a common ancestor. Rejection of this concept drives the efforts of creationists to impoverish the teaching of evolutionary biology in schools. It is hard to understand why, when so many very religious individuals and so many of the leaders of the major religious denominations find no problem accepting evolution along with their religion, this basic biological concept is so repugnant to creationists. The answer lies in the way humans think.

PATTERNS OF THOUGHT

It is useful, even though a simplification, to recognize that every one of us uses two patterns of thought. One pattern of thought relies on confirmable data and can be characterized as empirical or scientific and based on critical reasoning. The extraordinary advances in medical practice, technology, and agriculture are based on this mode of thought. The other pattern of thought involves romantic, religious, humanistic, or artistic elements. With this second pattern of thought we make decisions based on personal beliefs, preferences, desires, hopes, and faith.

Children employ both patterns of thought, but as they become older generally increase the rational pattern and decrease the romantic pattern. Nevertheless, numerous polls have shown that many adult Americans accept such supernatural notions as the ability to speak with the dead, the existence of ghosts, the coming of a comet to take true believers to heaven, or visits to the United States by aliens from space. Studies by psychologists have shown a strong tendency for the beliefs of childhood to remain throughout life. Even university courses designed to analyze the adequacy of the data supporting supernatural phenomena have only a slight effect in causing students to change their minds.

For example, Professor Thomas Gray reported the results of a study at Concordia University in Montreal in which 10 paranormal phenomena, such as astrology, UFOs, ghosts, and reincarnation were studied (1984). Students' belief in these phenomena was surveyed at the beginning of the course, its end, and a year later. The average percentage for belief in the 10 paranormal phenomena was 56 percent at the beginning of the course. After the phenomena were discussed in detail, the average decreased to 40 percent. When tested a year later, it had rebounded to 50 percent, increasing in 8 of the 10 cases. For two phenomena, astrology and ghosts, a higher level of belief was reported a year after the course ended than before the students had taken the course. The differences among the three tests were small, some not even significant, so the message is that a detailed study of the evidential basis of these 10 paranormal phenomena did almost nothing to deflate the level of belief.

In another study of belief systems, Singer and Benassi (1981, 49) note that the percentage of belief in the supernatural, such as ESP, may be as high as 80 to 90 percent in the general population. They note with concern that the level of such beliefs is increasing. "The current high level and strength of [supernatural] beliefs, which at least implicitly constitute a challenge to the validity of science and to the authority of the scientific community, may be a cause for our concern.... The psychological mechanisms involved in [supernatural] beliefs may represent more dramatic forms of some mundane pathologies of reasoning. To the extent that [supernatural] and mundane beliefs have similar determinants, the study of [supernatural] beliefs, which are at the outer limits of irrationality, may throw light on more ordinary reasoning pathologies." Then, after a long analysis, they conclude that, among other things, acceptance of the supernatural constitutes part of what it means to be human. This is not to be taken as an endorsement of the supernatural but as revealing "deficits in human inferences so universal and so stubborn that they can plausibly account for many of our errant beliefs, including [supernatural] ones. This recent spate of research, focusing on human problem-solving and judgmental processes, has resulted in an unflattering portrait of intuitive human reasoning" (Singer and Benassi, 1981, 50).

TEACHING SCIENCE AND TECHNOLOGY

An understanding of science and technology is so important for the welfare of human beings on Earth that a far greater effort must be made to produce a citizenry more competent in those twin disciplines than in the past. The solutions to many serious human problems will require the broad application of scientific knowledge—existing and yet to be discovered. The ultimate source of that knowledge is the community of scientists, and it is dangerous indeed to allow creationists to hinder that transmission of scientific knowledge to the general public and especially to schoolchildren.

The scientific community, however, has been remiss in not making certain that good science is taught in the schools, especially in elementary schools. In the elementary school years, youngsters are deeply interested in exploring their natural and human-made worlds and respond positively to opportunities to be deeply involved in that exploration. They should realize that natural phenomena have natural causes that can be discovered. It will take an extraordinary amount of wisdom, planning, and sympathetic understanding of the mind-set of parents and others to perfect a system of education that can accommodate these critical ways in which people think. Nevertheless it must be done not only for its importance in teaching the sciences in the schools, but also for the greater importance of allowing the adult generation to use rigorous thought based on state-of-the art thinking in all fields of science and, ultimately, on confirmable evidence.

Civilization is rapidly entering an era when decisions cannot be made on the basis of what we might like to happen but on scientific data and rational thought. If children can be exposed to intellectual problems from the earliest grades, by high school they should be able to deal explicitly with the evolution/creationism problem without being traumatized. The large number of individuals who are deeply religious and still accept science is proof that the human mind can employ both basic patterns of thought—each has its strengths and limitations, and each can be employed for its strengths. Therefore, a confirmable knowledge of the natural world is in the domain of science, but the meaning that we attach to our lives and to that of others is in the romantic domain. When teachers try to help children achieve this reconciliation of the two patterns of thought, they may find that parents are a problem in achieving this understanding, but time may solve that difficulty. After all, the children of today are the parents of tomorrow, and more understanding children will become more understanding parents.

Another touchy problem is how teachers present the evidential basis of science and creationism. Teachers should compare the evidential bases of evolution and creationism, pointing out that one base consists of confirmable evidence based on observations and experiments and the other on non-confirmable supernatural beliefs. One derives from the scientific way of reasoning, and the other from the romantic. It is unacceptable, of course, for a teacher to teach "creation science," that is, creationism as a science rather than a religious belief. This is not a matter of "academic freedom" any more that it would be for a science teacher to say 2 x 2 = 6, the value of ? is exactly 3, the Sun rotates around the Earth, or that there is good evidence for ghosts and the ability to speak with the dead. Science teachers should present the field as established by the professionals but also feel free, even encouraged, to deal with the unknown problems at the limits of knowledge where there is vigorous argument about the hypotheses being tested. Through such examples, students will see science in action and understand why ideas must be tested so rigorously before they can be incorporated into the corpus of accepted knowledge.

Such a consideration does not target the elimination of either the romantic or rational mode of thought but seeks ways of enhancing each. Much of the satisfaction and joy in our personal lives comes from the more romantic pattern, but the complex problems of the modern world demand the rigorous use of evidence, rational thought, and scientific approaches.

One simple change would greatly help to solve the problem—the creationists should cease telling the scientific community what should be taught in science classes. In a sense they should behave more like the scientists—I never heard of a scientist telling a preacher what to say from the pulpit on Sunday morning. The majority of Americans understand what is science and what is religion. They continue with their religious beliefs and look upon the natural world with greater understanding and awe. They understand the importance of science and religion in their lives and accept that both are critical for our life on Earth as we continue to grow in numbers, to pollute excessively, to exhaust natural resources, reduce biodiversity, and leave a world far less able to support a benign future for humanity.

REFERENCES

Gray, T. 1984. University course reduces belief in paranormal. *The Skeptical Inquirer* 8 (Spring): 247–251.

Singer, B., and V.A. Benassai. 1981. Occult beliefs. *American Scientist* 69 (January–February): 49–55.

AUTHOR'S NOTE

The National Academy of Sciences has published a booklet, *Teaching about Evolution and the Nature of Science* (1998), to help teachers deal with the evolution/creationism problem. It is available, free, online at *www.nap.edu/readingroom/books/evolution98*.

EDITOR'S NOTE

For more information about evolution, please see NSTA's Website devoted to evolution resources at *www.nsta.org/evresources.asp*.

EVOLUTION AND THE NATURE OF SCIENCE

A National Science Education Standards *perspective*

RODGER W. BYBEE

In the early 1990s, when work began on national standards for science education, those of us closely involved soon realized several of the limiting factors associated with the document we were about to produce. We knew the Standards would guide, but not control, the science education system. For example, the Standards would constitute national policies, not federal mandates. In the end, states and local jurisdictions would make decisions about the use of the Standards and the science curriculum. We also realized the National Standards would be controversial. That is, many states, groups, and individuals would simply resist the fact of standards being imposed on them, even if they were developed by a prestigious organization such as the National Research Council. Finally, we knew that some content—in particular evolution—would be controversial.

We also knew that national standards for science education would provide new possibilities for members of the science education community, especially science teachers. In general, the standards would give teachers the needed support for teaching science as inquiry, the mandate for continued professional development, an explanation of the essential role of assessment, clarification of definitive statements about science content, and an exploration of the importance of systemic support for the school science program.

I had a particular interest in the form and function that evolution and the nature of science would take in the Standards. In particular, the Standards had potential to provide a countervailing force in the controversy with various fundamental groups trying to eliminate evolution from the science curriculum. I knew that one national policy document would not end the assaults on the integrity of science and the professionalism of science teachers that ultimately affect the scientific literacy of students. I also knew that national standards could provide support for the teaching of evolution, if the content were scientifically accurate and educationally appropriate.

In this essay I review the perspective on evolution and the nature of science as presented in the *National Science Education Standards* (NRC, 1996). The discussion first presents the unifying concepts and processes and then proceeds to the content standards for evolution for grade levels K–4, 5–8, and 9–12. This is followed by a discussion of content outcomes for the nature of science using the same grade levels.

WHAT DO THE STANDARDS SAY ABOUT EVOLUTION?

The content standards include evolution in several different places, beginning with the standard for unifying concepts and processes for grades K–12. This standard provides the opportunity for students to integrate ideas from the natural and designed world. The concepts and processes of this standard are summarized in Figure 1. Note the explicit references to evidence, models, and explanation and evolution and equilibrium in this figure. The Standards continue with a statement on evolution that explains that: "Evolution is a series of changes, some gradual and some sporadic, that accounts for the present form and function of objects, organisms, and natural and designed systems. The general idea of evolution is that the present arises from materials and forms of the past. Although evolution is most commonly associated with the biological theory explaining the process of descent with modification of organisms from common ancestors, evolution also describes changes in the universe" (NRC, 1996, 119).

Thus, the unifying concepts and processes present evolution as a general process, for which biological evolution is a specific case. Other specific cases might include stellar evolution and evolution of the atmosphere.

The original intent of this unifying standard was to provide an organizing framework for those curriculum developers, states, and school districts with intentions to design and implement integrated approaches to science. To the degree those groups would base their programs on the national standards, we wished to ensure the inclusion of evolution and the nature of science. The remaining content standards use grade levels and disciplines as the organizer. The content standards for science are divided into grade levels of K–4, 5–8, and 9–12. The content standards used physical science, life science, and Earth and space science as organizers for the disciplines.

Grades K–4: Evolution

At grades K–4, the fundamental concepts and principles in the life sciences are organized by the following categories: The Characteristics of Organisms, Life Cycles of Organisms, and Organisms and Their Environments. Although evolution is not explicitly mentioned at the K–4 level, the content standards are intended to develop scientific knowledge and understanding as a foundation for the concepts introduced in middle and high school.

FIGURE 1. Unifying Concepts and Processes

As a result of activities in grades K–12, all students should develop understanding and abilities aligned with the following concepts and processes:

- Systems, order, and organization
- **Evidence, models, and explanation**
- Constancy, change, and measurement
- **Evolution and equilibrium**
- Form and function

National Research Council. 1996. *National Science Education Standards.* Washington, DC: National Academy Press.

During the elementary grades, children should build their understanding of biological concepts through direct experience with organisms, their life cycles, and their habitats. These experiences emerge from the sense of wonder and natural interests of children who ask questions such as: "How do plants get food?" "How many different animals are there?" "Why do some animals eat other animals?" "What is the largest plant?" "Where did the dinosaurs go?" An understanding of the characteristics of organisms, life cycles of organisms, and the complex interactions among all components of the natural environment begins with questions such as these and an understanding of how individual organisms maintain and continue life. With an educational approach such as this, students should be ready for the middle years of school.

Grades 5–8: Evolution

The life science standard for grades 5–8 includes the category of Diversity and Adaptations of Organisms. The fundamental concepts of this category are described in Figure 2. Middle school is the first level at which biological evolution is specifically addressed by name.

Often, adolescent students and many adults do not understand evolution, in particular the connection between genetic variations in a species, the process of selection, and the idea that evolution occurs in populations. Many students hold the Lamarckian idea that there is a direct link between adaptation of an organism and environmental conditions, without fully understanding the gradual process of selection for genetic variation within the species.

The Standards discuss the difficulty teachers may encounter. "Understanding adaptation can be particularly troublesome at this level. Many students think adaptation means that individuals change in major ways in response to environmental changes (that is, if the environment changes, individual organisms deliberately adapt)" (NRC, 1996, 156).

The content standards for Earth science, grades 5–8, also introduce evolution. The discussion of Earth's history describes linkages between the Earth's history and the history of life. Here is a quote from the standard on Earth's history: "The earth processes we see today, including erosion, movement of lithospheric plates, and changes in atmospheric composition, are similar to those that occurred in the past. Earth's history is also influenced by occasional catastrophes, such as the impact of an asteroid or comet.

Fossils provide important evidence of how life and environmental conditions have changed" (NRC, 1996, 160).

The study of Earth's history in the middle grades should present students with some evidence for the co-evolution of major features and domains, such as distribution of continents, landforms, composition of the atmosphere, global climate, and biodiversity.

Grades 9–12: Evolution

The life science standards for grades 9–12 directly address biological evolution. The Earth science standards also introduce evolution. Figures 3 and 4 display the standards and the fundamental concepts that science teachers should use as they provide opportunities for students to learn more accurate scientific understandings about evolution.

It is obvious that the Standards present evolution as a major concept. Beginning in the elementary grades, evolution is fundamental to science education.

WHAT DO THE STANDARDS SAY ABOUT THE NATURE OF SCIENCE?

In the introduction to this book, I made the point that science teachers have the opportunity, even the obligation, to help students develop a better understanding of the nature of scientific knowledge. To this end, the Standards also provide guidance and support for teachers.

SECTION II — THE EDUCATIONAL PERSPECTIVE

Grades K–4: Nature of Science

Relative to the nature of science, the main message of the Standards for elementary grades centers on experiences that develop abilities and understandings of inquiry and the realization that science is a human endeavor. The Standards state: "From the earliest grades, students should experience science in a form that engages them in the active construction of ideas and explanations and enhances their opportunities to develop the abilities of doing science. Teaching science as inquiry provides teachers with the opportunity to develop student abilities and to enrich their understanding of science" (NRC, 1996, 121).

Activities that have children ask questions about the natural world, plan simple investigations, use equipment and tools to gather data, use data to form explanations (i.e., an-

---

Excerpt from Life Science Content Standard Grades 5–8

As a result of their activities in grades 5–8, all students should develop an understanding of

- Structure and function in living systems
- Reproduction and heredity
- **Regulation and behavior**
- Populations and ecosystems
- **Diversity and adaptations of organisms**

Fundamental concepts and principles that underlie this standard include

Regulation and Behavior

- An organism's behavior evolves through adaptation to its environment. How a species moves, obtains food, reproduces, and responds to danger are based in the species' evolutionary history.

Diversity and Adaptations of Organisms

- Millions of species of animals, plants, and microorganisms are alive today. Although different species might look dissimilar, the unity among organisms becomes apparent from an analysis of internal structures, the similarity of their chemical processes, and the evidence of common ancestry.

- Biological evolution accounts for the diversity of species developed through gradual processes over many generations. Species acquire many of their unique characteristics through biological adaptations, which involves the selection of naturally occurring variations in populations. Biological adaptations include changes in structures, behaviors, or physiology that enhance survival and reproductive success in a particular environment.

- Extinction of a species occurs when the environment changes and the adaptive characteristics of a species are insufficient to allow its survival. Fossils indicate that many organisms that lived long ago are extinct. Extinction of species is common; most of the species that have lived on the earth no longer exist.

National Research Council. 1996. *National Science Education Standards.* Washington, DC: National Academy Press.

swers to their questions), and communicate their investigations and explanations all set a foundation for understanding inquiry and the nature of science.

Grades 5–8: Nature of Science

Figure 5 displays excerpts of content from the standard on the history and nature of science. The content described here will help science teachers make decisions about the curriculum and what students should learn.

Grades 9–12: Nature of Science

High school students should build on the concepts developed in earlier years and develop a reasonable understanding of science as a distinctive way of knowing and explaining the natural world. Figure 6 provides content that will inform science teachers' decisions about what students should know about the nature of science.

I should note that in addition to understanding evolution in the context of the nature of science, students should also learn about it from a

---

FIGURE 3: Excerpt from Life Science Content Standards, Grades 9–12

As a result of their activities in grades 9–12, all students should develop an understanding of

- The cell
- Molecular basis of heredity
- **Biological evolution**
- Interdependence of organisms
- Matter, energy, and organization in living systems
- Behavior of organisms

Fundamental concepts and principles that underlie this standard include

Biological Evolution

- Species evolve over time. Evolution is the consequence of the interactions of (1) the potential for a species to increase its numbers, (2) the genetic variability of offspring due to mutation and recombination of genes, (3) a finite supply of the resources required for life, and (4) the ensuing selection by the environment of those offspring better able to survive and leave offspring.

- The great diversity of organisms is the result of more than 3.5 billion years of evolution that has filled every available niche with life forms.

- Natural selection and its evolutionary consequences provide a scientific explanation for the fossil record of ancient life forms, as well as for the striking molecular similarities observed among the diverse species of living organisms.

- The millions of different species of plants, animals, and microorganisms that live on earth today are related by descent from common ancestors.

- Biological classifications are based on how organisms are related. Organisms are classified into a hierarchy of groups and subgroups based on similarities which reflect their evolutionary relationships. Species is the most fundamental unit of classification.

National Research Council. 1996. *National Science Education Standards.* Washington, DC: National Academy Press.

Excerpt from Earth Science Content Standards, Grade 9–12

As a result of their activities in grades 9–12, all students should develop an understanding of

- Energy in the earth system
- Geochemical cycles
- **Origin and evolution of the earth system**
- **Origin and evolution of the universe**

Fundamental concepts and principles that underlie this standard include

The Origin and Evolution of the Earth System

• The sun, the Earth, and the rest of the solar system formed from a nebular cloud of dust and gas 4.6 billion years ago. The early earth was very different from the planet we live on today.

• Geologic time can be estimated by observing rock sequences and using fossils to correlate the sequences at various locations. Current methods include using the known decay rates of radioactive isotopes present in rocks to measure the time since the rock was formed.

• Interactions among the solid earth, the oceans, the atmosphere, and organisms have resulted in the ongoing evolution of the earth system. We can observe some changes such as earthquakes and volcanic eruptions on a human time scale, but many processes such as mountain building and plate movements take place over hundreds of million of years.

• Evidence for one-celled forms of life—the bacteria—extends back more than 3.5 billion years. The evolution of life caused dramatic changes in the composition of the earth's atmosphere, which did not originally contain oxygen.

The Origin and Evolution of the Universe

• The origin of the universe remains one of the greatest questions in science. The "big bang" theory places the origin between 10 and 20 billion years ago, when the universe began in a hot dense state; according to this theory, the universe has been expanding ever since.

• Early in the history of the universe, matter, primarily the light atoms hydrogen and helium, clumped together by gravitational attraction to form countless trillions of stars. Billions of galaxies, each of which is a gravitationally bound cluster of billions of stars, now form most of the visible mass in the universe.

• Stars produce energy from nuclear reactions, primarily the fusion of hydrogen to form helium. These and other processes in stars have led to the formation of all the other elements.

National Research Council. 1996. *National Science Education Standards.* Washington, DC: National Academy Press.

historical perspective—that is, as one of the important advances in science.

CRITERIA FOR DEVELOPING AND USING STANDARDS

The majority of states and school districts are developing standards and most have used the National Science Education Standards as a resource. Those individuals who have the task of developing standards need to keep in mind several fundamental criteria that guided the development of the Standards. Based on obligations to science, students, and schoolteachers, I recommend the following three criteria for the selection of content.

Figure 5. Excerpts from History and Nature of Science, Grades 5–8

As a result of activities in grades 5–8, all students should develop an understanding of

- Science as a human endeavor
- **Nature of science**
- History of science

Fundamental concepts and principles that underlie this standard include

Nature of Science

- Scientists formulate and test their explanations of nature using observation, experiments, and theoretical and mathematical models. Although all scientific ideas are tentative and subject to change and improvement in principle, for most major ideas in science, there is much experimental and observational confirmation. Those ideas are not likely to change greatly in the future. Scientists do and have changed their ideas about nature when they encounter new experimental evidence that does not match their existing explanations.

- In areas where active research is being pursued and in which there is not a great deal of experimental or observational evidence and understanding, it is normal for scientists to differ with one another about the interpretation of the evidence or theory being considered. Different scientists might publish conflicting experimental results or might draw different conclusions from the same data. Ideally, scientists acknowledge such conflict and work towards finding evidence that will resolve their disagreement.

- It is part of scientific inquiry to evaluate the results of scientific investigations, experiments, observations, theoretical models, and the explanations proposed by other scientists. Evaluation includes reviewing the experimental procedures, examining the evidence, identifying faulty reasoning, pointing out statements that go beyond the evidence, and suggesting alternative explanations for the same observations. Although scientists may disagree about explanations of phenomena, about interpretations of data, or about the value of rival theories, they do agree that questioning, response to criticism, and open communication are integral to the process of science. As scientific knowledge evolves, major disagreements are eventually resolved through such interactions between scientists.

National Research Council. 1996. *National Science Education Standards.* Washington, DC: National Academy Press.

First, the subject matter of science should be presented as accurately as possible in the overall conceptualization for a discipline, such as biology. This obligation means that the teacher maintains the integrity of biology by indicating that biological evolution is the central organizing idea for the discipline.

Second, the teacher should present the subject matter of science in developmentally appropriate forms; that is, the concepts and abilities should be aligned with the students' ages and stages of development.

Finally, the obligation to science teachers requires that those who develop or use science standards state as clearly and unambiguously as possible the fundamental concepts of evolution and the nature of science. Some will see this as a cause for concern, if not outright conflict,

---

FIGURE 6: Excerpt from History and Nature of Science Standards, Grades 9–12

As a result of activities in grades 9–12, all students should develop understanding of

- Science as a human endeavor
- **Nature of scientific knowledge**
- Historical perspectives

Fundamental concepts and principles that underlie this standard include

Nature of Scientific Knowledge

- Science distinguishes itself from other ways of knowing and from other bodies of knowledge through the use of empirical standards, logical arguments, and skepticism, as scientists strive for the best possible explanations about the natural world.

- Scientific explanations must meet certain criteria. First and foremost, they must be consistent with experimental and observational evidence about nature and must make accurate predictions, when appropriate, about systems being studied. They should also be logical, respect the rules of evidence, be open to criticism, report methods and procedures, and make knowledge public. Explanations on how the natural world changes based on myths, personal beliefs, religious values, mystical inspiration, superstition, or authority may be personally useful and socially relevant, but they are not scientific.

- Because all scientific ideas depend on experimental and observational confirmation, all scientific knowledge is, in principle, subject to change as new evidence becomes available. The core ideas of science such as the conservation of energy or the laws of motion have been subjected to a wide variety of confirmations and are therefore unlikely to change in the areas in which they have been tested. In areas where data or understanding are incomplete, such as the details of human evolution or questions surrounding global warming, new data may well lead to changes in current ideas or resolve current conflicts. In situations where information is still fragmentary, it is normal for scientific ideas to be incomplete, but this is also where the opportunity for making advances may be greatest.

National Research Council. 1996. *National Science Education Standards.* Washington, DC: National Academy Press.

between biology educators and those who oppose teaching evolution. I encourage another view. Science educators can use the accuracy, appropriateness, and authority of the Standards to support their position in debates with those who want to introduce nonscientific topics and ideas in science classrooms. Science teachers can use the Standards and the fact that they have the imprimatur of the National Research Council to support accurate science and exemplary science teaching.

The teams of scientists, science educators, and science teachers who developed the Standards thought that it would be important to address the ways and means of using standards. Relative to this article, I note the following:

- All eight categories of content standards should be included in the science curriculum. For example, students should have opportunities to learn science in personal and social perspectives and to learn about the history and nature of science, as well as learning life science subject matter.
- In a science program, all standards from a category should be addressed. For instance, "biological evolution" should not be eliminated from the life science standards.
- Science content can be added. The connections, depth, details, and selection of topics can be enriched and varied as appropriate. Of course, addition of content must not prevent students from learning fundamental concepts.

For the science education community, the first statement supports the inclusion of historical study of Darwin's theory and the development of that theory since Darwin's formulation. This history also provides the opportunity for students to learn about the nature of science. The second point, including the example, was made with the express purpose of countering those who would eliminate biological evolution or state it in some other diluted form. The third point is meant to show that the Standards do not represent the *only* content for school science programs. For example, a school science program might include advanced placement science courses or honors seminars.

BEYOND STANDARDS

Since release of the Standards in 1995, the National Academy of Science has recognized the need to support the work of science teachers, especially in the teaching of evolution. In 1998 the National Academy of Sciences released *Teaching about Evolution and the Nature of Science* (NAS, 1998). This small volume provides science teachers with a rationale for teaching about evolution, an update on major themes in evolution, and an introduction to the nature of science presented in the context of evolution. In addition, it provides several classroom activities and answers to frequently asked questions about evolution. In 1999 the National Academies published the second edition of *Science and Creationism* (NAS, 1999), a valuable resource for science teachers.

CONCLUSION

The National Academy of Sciences and the National Research Council developed the Standards with the clear intention of ensuring the best science education for all students and supporting science teachers in achieving this goal. Science teachers should be familiar with the idea that a new medicine selects for its own demise. So it is with the various new strategies used by the scientific, educational, and legal communities to counter creationists (AAAS, 1996). We can expect the debate between creationists and the scientific community to continue, with the likely field of these battles being school science programs. The *National Science Education Standards* and *Teaching about Evolution and the Nature of Science* present constructive approaches to teaching about evolution and the nature of science.

REFERENCES

American Association for the Advancement of Science. 1996. Creationists evolve new strategy. *Science News and Views* 273 (July 26): 420–422.

National Academy of Sciences. 1998. *Teaching about Evolution and the Nature of Science.* Washington, DC: National Academy Press.

National Academy of Sciences. 1999. *Science and Creationism: A View from the National Academy of Sciences.* Washington, DC: National Academy Press.

National Research Council. 1996. *National science education standards.* Washington, DC: National Academy Press.

THE EDUCATIONAL PERSPECTIVE

SECTION II

DO STANDARDS MATTER?

How the quality of state standards relates to evolution instruction

RANDY MOORE

In recent years, more and more U.S. states have drafted standards designed to promote quality science teaching. Although most states include evolution instruction in their standards, the teaching of evolution in public schools remains problematic. To what extent is the weak treatment of evolution in public schools a manifestation of weak science education standards?

In *Good Science, Bad Science: Teaching Evolution in the States,* California State University physicist and science curriculum specialist Lawrence Lerner (2000) evaluated the quality of states' standards for teaching evolution. Lerner's widely publicized report made a variety of disappointing conclusions. The following are some examples:

- 10 states received an A, indicating very good to excellent treatment of evolution in their educational standards;
- 14 states received a B, indicating good treatment of evolution in their educational standards;
- 7 states received a C, indicating satisfactory treatment of evolution in their educational standards;
- 6 states received a D, indicating unsatisfactory treatment of evolution in their educational standards;
- 13 states received an F or F–, indicating that their standards are "useless for purposes of teaching evolution." Ten of the states that received grades

of D or worse do not use the word *evolution* in their educational guidelines, and one uses the word only once.

State educational standards are supposed to be the foundation of what students learn to produce the state's desired educational outcomes. How do state standards for teaching evolution relate to the attitudes and actions of biology teachers?

EVOLUTION: THE MISSING SUBJECT

To answer this question, I studied how Lerner's evaluations of state guidelines for teaching evolution correlate with various evolution-related policies and characteristics of biology teachers in public schools. I was not surprised to learn that states whose standards received the lowest grades for teaching evolution have relatively large percentages of biology teachers who question the validity of evolution, want to teach creationism, or actively teach creationism. For example, at least one-third of biology teachers in Kentucky, Ohio, Illinois, Kansas, and Georgia (all of which received a D or F) want creationism to be taught in science classes (Aguillard, 1999; Moore, 1999; Zimmerman, 1987; Nickels and Drummond, 1985, Buckner, 1983; Aldrich, 1991; Ellis, 1986). In fact, many teachers in these states actually do teach creationism, even though *Edwards v. Aguillard* established that the teaching of creationism in public school science classes is unconstitutional (Moore, 2000).

Similarly, in Louisiana—a state having C standards—29 percent of biology teachers want to teach creationism in their classes, and 41 percent of biology teachers believe that creationism may be a scientifically valid concept. Almost one-fourth of biology teachers in Louisiana put little or no emphasis on evolution in their courses, and many of Louisiana's biology teachers don't recall hearing the word evolution in their biology courses (Aguillard, 1999). States that received low grades for teaching evolution frequently embrace anti-science policies, such as disclaimers in biology textbooks in Alabama and a Kentucky law that encourages teachers to give "equal time" to creationism.

I was surprised, however, to find a similar situation in states having "good" and "excellent" standards for teaching evolution. For example, in Indiana, a state having A standards, 43 percent of biology teachers characterize their teaching of evolution as "avoidance" or "briefly mention," and at least 20 percent of biology teachers do not accept or are undecided about the scientific validity of evolution (Rutledge and Mitchell, 2002). Similarly in Pennsylvania, another state having A standards, one third of biology teachers do not believe that evolution is central to biology (Weld and McNew, 1999). In Oregon and South Dakota, states having B standards, 26 to 39 percent of biology teachers want to teach creationism in their courses (Weld and McNew, 1999; Affanato, 1986), and in Minnesota, another state having B standards, 40 percent of biology teachers spend little or no time teaching evolution (Hessler, 2000).

Clearly, state standards for teaching evolution are unrelated to the actual teaching of evolution in biology classrooms. As Louisiana biology teacher Don Aguillard concluded after his study of the attitudes and actions of biology teachers in Louisiana, "creationism is alive and well among biology teachers" (Moore, 1999, 172). In light of this, it's easy to understand how "over a quarter—and perhaps as many as half—of the nation's high school students get educations shaped by creationist influence—in spite of the overwhelming opposition of the nation's scientific, educational, intellectual, and media establishments" (Eve and Harrold, 1991, 167).

Of course, state standards alone do not determine what is taught in U.S. schools; what students learn about evolution is also influenced by tests, the curriculum, and textbooks (Lerner, 2000). Lerner wondered to what extent the poor teaching of evolution is due to weak standards and to what extent it reflects something more complicated. One of the more complicated answers is, in fact, relatively simple: many biology teachers avoid teaching evolution, endorse creationism, or—in some cases—teach creationism.

NOT MUCH HAS CHANGED

Various professional organizations and prominent biologists have called for high standards for the teaching of evolution (Grobman, 1998). However, these standards have made little difference. For example, in 1941, biologist Oscar Riddle noted that creationism was popular among biology teachers and that fewer than half of high school biology teachers taught evolution (1941). In 1959, geneticist and Nobel laureate Hermann Muller again noted the popularity of creationism among biology teachers when stating that biology teaching was dominated by "antiquated religious traditions" (1959). Similarly, when the National Association of Biology Teachers (NABT) established its "Fund for Freedom in Science Teaching" in the 1970s to help teachers resist the anti-science activities of creationists, many members wrote to demand that NABT give creationists equal time and stop trying to "silence people," "persecute creationists," and "promote atheism and agnosticism in the schools" (Nelkin, 1982). In response, NABT sponsored a session about creationism at its annual meeting and published several pro-creationism articles in its journal, *The American Biology Teacher* (Gish, 1970; Moore, 1973). In the year 2000, NABT recognized a creationist with one of its Outstanding Biology Teacher Awards. Today, large percentages of biology teachers continue to endorse creationism and reject evolution (Harp, 1999; Moore, 2001b).

If we're to actually improve how evolution is taught, we must do more than establish standards that are often ignored. Indeed, a variety of organizations have emphasized that students should have a thorough understanding of evolution (American Association for the Advancement of Science, 1989; National Association of Biology Teachers, 1997; National Academy of Science, 1998; National Research Council, 1985; National Science Teachers Association, 1997), but neither science education programs nor biology teachers have been held accountable for teaching evolution.

We'd never consider hiring a science teacher who questions or rejects the basic laws of motion, or a chemist who questions or rejects the basic laws of chemistry. Yet we regularly hire, retain, and reward biology teachers who question or reject evolution, and who want to replace basic laws of science with religion. Why would we expect to have an effective science education program when teachers themselves question or reject basic principles of science?

If we're ever to begin remedying the dismal status of evolution education in the United States, we must insist on rigorous standards for the teaching of evolution *and implement those standards*. We must hire, retain, and reward science teachers who openly endorse and effectively teach science, not creationism.

ONLINE EXTENSION

For further data on state standards and evolution, NSTA members can log on to *TST* at *www.nsta.org/highschool*.

REFERENCES

Affannato, F.E. 1986. *A survey of biology teachers' opinions about the teaching of evolutionary theory and/or the creation model in the United States in public and private schools.* Unpublished Ph.D. dissertation, University of Iowa.

Aguillard, D. 1999. Evolution education in Louisiana public schools: A decade following Edwards v. Aguillard. *The American Biology Teacher* 61(3): 182–188.

Aldrich, K.J. 1991. Teachers' attitudes toward evolution and creationism in Kansas biology classrooms. *Kansas Biology Teacher* 8(1): 20–21.

American Association for the Advancement of

Science. *Project 2061.* 1989. *Science for all Americans.* Washington, D.C.: American Association for the Advancement of Science.

Buckner, E.M. 1983. *Professional and political socialization: High school science teacher attitudes on curriculum decisions, in the context of the "scientific" creationism campaign.* Ph.D. dissertation. Georgia State University. Ann Arbor, Mich.: University Microfilms International.

Ellis, W.E. 1986. Creationism in Kentucky: The response of high school biology teachers. In *Science and Creation.* R.W. Hanson, ed. New York: Macmillan. 72–91.

Eve, R., and F. Harrold. 1991. *The Creationist Movement in Modern America.* Boston: Twayne.

Gish, D. 1970. A challenge to neo-Darwinism. *The American Biology Teacher* 32: 95–96.

Grobman, A. 1998. National standards. *The American Biology Teacher* 60(8): 4.

Harp, L. 1999. School guide drops word "evolution." *Louisville Courier–Journal* (5 October), pp. Al, A5.

Hessler, E. 2000. Two "e" words: ecology and evolution. *Minnesota Science Teachers Association Newsletter* 37(2): 6.

Lerner, L.S. 2000. *Good Science, Bad Science: Teaching Evolution in the States.* Washington, D.C.: The Thomas B. Fordham Foundation. Online at www.edexcellence.net.

Moore, J.N. 1973. Evolution, creationism, and the scientific method. *The American Biology Teacher* 35: 23–26.

Moore, R. 1999. The courage and convictions of Don Aguillard. *The American Biology Teacher* 61(3): 166–174.

Moore, R. 2001a. Educational malpractice: Why do so many biology teachers endorse creationism? *Skeptical Inquirer* 25(6): 38–43.

Moore, R. 2001b. *Evolution in the Courtroom: A Reference Guide.* Denver, Colo.: ABC-CLIO Publishers.

Muller, H.J. 1959. One hundred years without Darwinism are enough. *The Humanist* 19: 139.

National Academy of Sciences. 1998. *Teaching About Evolution and the Nature of Science.* Washington, D.C.: National Academy Press.

National Association of Biology Teachers. 1997. Position statement on the teaching of evolution. *NABT News and Views* (June), 4–5.

National Research Council. 1985. *Mathematics, Science and Technology Education: A Research Agenda.* Washington, D.C.: National Academy Press.

National Science Teachers Association. 1997. An NSTA position statement on the teaching of evolution. *Journal of College Science Teaching* 27(1): 7–8.

Nelkin, D. 1982. *The Creation Controversy: Science or Scripture in the Schools?* New York: Norton.

Nickels, M.K. and B.A. Drummond. 1985. Creation/evolution: Results of a survey conducted at the 1983 ISTA convention. *Creation/Evolution Newsletter* 5(6): 2–15.

Riddle, O. 1941. Preliminary impressions and facts from a questionnaire on secondary school biology. *The American Biology Teacher* 3: 151–159.

Rutledge, M.L. and M.A. Mitchell. 2002. High school biology teachers' knowledge structure, acceptance, and teaching of evolution. *The American Biology Teacher* 64: 21–28.

Weld, J. and J.C. McNew. 1999. Attitudes toward evolution. *The Science Teacher* 66(9): 27–31.

Zimmerman, M. 1987. The evolution-creation controversy: Opinions of Ohio high school biology teachers. *Ohio Journal of Science* 87: 115–125.

EVOLUTION: DON'T DEBATE, EDUCATE

Teach inquiry and the nature of science

RODGER W. BYBEE

From time to time prudent science teachers pause and reflect on problems that confront the profession. Specifically, I am referring to the continuing controversy over the teaching of biological evolution and other scientific ideas such as the Big Bang theory. Teachers and others in the science community have concerns about nonscientific ideas being introduced into the school curriculum. Even though evidence against creationism is substantial (Pennock, 1999; Eldridge, 2000), for more than 100 years we have been unsuccessful in countering this ideology. Past strategies have been primarily to criticize the inconsistency of creationists' thinking and the general view that their position is dogmatic and, by definition, not open to scientific attitudes such as doubt or skeptical review (Moore, 2000). Reflecting on a reasonable response to this controversy brings to mind a quotation from Thomas Jefferson and the idea behind the theme of this article. He said, "I know of no safe depository of the ultimate powers of the society but the people themselves; and if we think them not enlightened enough to exercise their control with a wholesome discretion, the remedy is not [to] take it from them, but to inform their discretion" (Baron, 1993, 28). According to my interpretation and application of Jefferson's idea, teachers should inform their students about the nature of science so, in time, science education will find greater support and understanding from the general public. Although controversies will undoubtedly continue, scientists should balance their desire for political, legal, and educational confrontation with creationists with the long-term goal of educating all students about science.

So, concerning the continuing controversy with creationists, how should science teachers proceed? I have two recommendations. First, they should avoid debating creationists. Second, they should help students develop a greater understanding and appreciation for science as a way of explaining the natural world (Moore, 1993; National Academy of Sciences, 1998; Weld and McNew, 1999). Emphasizing inquiry and the nature of science strengthens the public's understanding of science and provides an appropriate position for all science teachers. Without reducing the role and place of evolution in the curriculum, we should increase our emphasis on developing students' understanding of the nature of science and their science inquiry abilities.

AVOID DEBATING

Debating creationists may not be the best strategy for most science teachers, but some individuals, groups, and organizations such as the National Center for Science Education must confront and debate those attempting to assert their control over the science curriculum. To completely ignore even the smallest assault on the integrity of science opens the way for misconceptions about the role of science in individual lives and our society. Such incursions confuse the purposes and importance of both religion and science. Although encounters such as one parent suggesting equal time for creationist views or a school board member advocating disclaimers in textbooks may seem small and easy to ignore, the price of scientific integrity is high and requires a clear, direct, and strong response to even these apparently small assaults. Just as fundamentalist groups advocating nonscientific views are organized and have well-developed strategies, so too, should those who confront them be knowledgeable of effective approaches and strategies. The majority of science teachers, however, should avoid debating creationists. They should strive for the larger and longer-term benefits of teaching all students about science and placing confidence in the ultimate powers of society and citizens enlightened about inquiry and the nature of science.

There are several reasons not to debate creationists. First, to put it simply, most scientists, science educators, and science teachers probably will lose the debate *in the eyes of the public*. Teachers are not prepared for the questions and issues that will be front and center. For example, the debate often is not about evolutionary concepts and processes; it is about politics and power—essentially, control of the science curriculum. If science teachers lose the debate, they give political power to creationists, regardless of the scientific accuracy of the teachers' positions, or the inaccuracy of the creationists' position.

> Although controversies will undoubtedly continue, scientists should balance their desire for political, legal, and educational confrontation with creationists with the long-term goal of educating all students about science.

My second point comes from the book, *Rocks of Ages* (Gould, 1999), in which the term *magisterium* is used to describe a domain that holds the appropriate tools for meaningful discourse and resolution of issues. The magisterium of science addresses issues empirically and proposes explanations for what the natural world is made of and how it works. For science, the goal of this domain is summarized by the term *empirical truth*. The magisterium of religion is different, though, as it addresses issues of ultimate meaning and moral values. For religion, the goal is spiritual understanding and goodness. To use an everyday characterization, science studies how the heavens go, and religion studies how to go to heaven. These two magisteria do not overlap, they need not be in conflict, and neither one addresses all ways of knowing. For example, neither encroaches on the magisterium of the arts and humanities, which present other ways of knowing, specifically about beauty and aesthetics.

Science teachers should be aware of the official scientific positions offered by professional organizations such as the National Science Teachers Association (1997) and the National Academy of Sciences (1999). These documents describe the magisterium of science. Unfortunately, we live in an age when people, and especially the media, have difficulty reconciling the fact that science and religion use independent and different ways of addressing issues. So the media and popular culture suggest that we must give equal time, we must pit science against religion, and one magisteria must win and the other lose. Alternatively, some argue that both science and religion represent the same quest, so they should be integrated in the school curricula (Nord, 1999). Debating science and religion lends credence to the misconception that one magisteria holds dominance over the other, that one must win and one must lose. Ultimately, debating does neither science nor religion any good (Ayala, 2000).

The third reason not to debate is that debates about evolution and creationism are rarely about evolution. The debates often hinge on issues related to inquiry and the nature of science. Those supporting the scientific side are confronted with myriad questions about how science works: What is a theory? What is the difference between theory and fact? What serves as proof of a theory? What is the role of authority versus evidence? What counts as evidence in support of an explanation? How do scientific theories change?

In actual debates, such questions are often not posed as questions; they are stated as assertions such as, "Science does not have facts, only theories." Or, "Scientists have not proven evolution." The debate format gives little or no time for an elaborate answer. This leaves the audience with a view that the assertions hold merit; thus the public has the perception that the scientific responses were inadequate. Therefore, debates are not really about clarifying issues, learning others' positions, or having an educational dialogue. From the

> Without reducing the role and place of evolution in the curriculum, we should increase our emphasis on developing students' understanding of the nature of science and their science inquiry abilities.

creationists' view, they are about winning, about power over the science curriculum, and about leaving doubt in the minds of those who may not have a clearly defined position about what should and should not be in school science programs.

My final point about debating is one that most science teachers who have debated will confirm—the process is usually unrewarding and often not civil. The encounter is not like classroom experiences of teaching about evolution and the nature of science where one can experience the joy and excitement of students' learning more about their world. Debating creationists often leaves one with a sense of inadequacy, frustration, and defeat.

As science teachers, debating creationists is not our calling, it is not our purpose, and it is not in our professional interest. To the degree teachers help students recognize the difference between science and nonscience, the tools for meaningful scientific discourse, and the resolution of differences between scientific explanations, we can inform the public's discretion and enhance the support for science and science education.

INQUIRY, THE NATURE OF SCIENCE, AND EVOLUTION

In the late sixteenth and early seventeenth centuries, the emergence of modern science was based not so much on new knowledge about the world as it was on new ways of thinking about the world. Simply stated, a scientific way of knowing about the natural world is based on evidence from observations and experiments. Further, scientific understanding must build on extant knowledge and be tested against nature. Scientific explanations are subject to the challenges of empirical evidence. Before modern science, explanations for the way things worked were based on authorities such as Aristotle or on religious dogma. Scientific progress required the combination of empirical evidence from nature and human reasoning and imagination.

The combination of evidence, reasoning, and imagination develops through a variety of forms. We might think of this synergy as originating with a question about nature, followed by the design of specific ways to collect evidence to answer the question, and finally the formulation and skeptical review of a proposed answer to the question. Regardless of form, acquiring evidence eventually gives rise to explanations constructed through human understanding and imagination. These explanations, or theories, are the fullest answers to questions about nature. Although scientists make no claims about ultimate truth, scientists do continually strive to formulate more consistent and coherent theories of nature.

Teaching about the nature of science necessitates teaching about theories such as biological evolution—something we do not do well in science education (Lerner and Bennetta, 1988). The formulation, testing, and development of theories are among the central and most important activities of the scientific community. Further, generally accepted theories have a logical structure for established knowledge (Lewis, 1986). The logical structure of knowledge is an educational connection presented for science education more than 40 years ago in *The Process of Education* (Bruner, 1960).

Teaching about the nature of science should be integrated with teaching about evolution. Inquiry and the nature of science are not entities separate from the development of scientific theories. One only has to review books such as *On the Origin of Species by Means of Natural Selection* (Darwin, 1859), *One Long Argument* (Mayr, 1991), or *Science as a Way of Knowing* (Moore, 1993) to see the relationships between the scientific processes and the structure and development of a theory such as biological evolution. In short, developing an understanding of the structure of a theory provides learners with intellectual structures and logical associations that facilitate learning.

Although the view of teaching science presented here may seem an obvious element of education, students are seldom asked to reflect on the nature of current scientific knowledge and the means by which scientists come to know what they do about nature. We can confirm this observation by reviewing almost any commonly used science textbook. For the most part, science textbooks present science as a body of knowledge, a "rhetoric of conclusion," not as a way of knowing. As such, our textbooks and teaching present scientific knowledge in an authoritative manner, one that represents pre-seventeenth century perspectives.

When we do teach about the nature of science, it is often presented as a systematic method or a number of disarticulated processes. On the one hand, the method perspective presents students with the idea that when, and only when, scientists follow a set number of steps to build new knowledge, science will progress. On the other hand, the process involves teaching students such skills as observing, hypothesizing, and inferring. This often leaves students with little or no concept of how all the processes work together to result in new or improved scientific explanations.

Nature of Scientific Knowledge: Fundamental Concepts and Principles

As a result of activities in grades 9–12, all students should develop understandings of

- Science as a human endeavor
- Nature of scientific knowledge
- Historical perspectives

Fundamental concepts and principles that underlie this standard include:

- Science distinguishes itself from other ways of knowing and from other bodies of knowledge through the use of empirical standards, logical arguments, and skepticism, as scientists strive for the best possible explanations about the natural world.

- Scientific explanations must meet certain criteria. First and foremost, they must be consistent with experimental and observational evidence about nature and must make accurate predictions, when appropriate, about systems being studied. They should also be logical, respect the rules of evidence, be open to criticism, report methods and procedures, and make knowledge public. Explanations about how the natural world changes based on myths, personal beliefs, religious values, mystical inspiration, superstition, or authority may be personally useful and socially relevant, but they are not scientific.

- Because all scientific ideas depend on experimental and observational confirmation, all scientific knowledge is, in principle, subject to change as new evidence becomes available. The core ideas of science, such as the conservation of energy or the laws of motion, have been subjected to a wide variety of confirmations and are therefore unlikely to change in the areas in which they have been tested. In areas where data or understanding are incomplete, such as the details of human evolution or questions surrounding global warming, new data may well lead to changes in current ideas or resolve current conflicts. In situations where information is still fragmentary, it is normal for scientific ideas to be incomplete, but this is also where the opportunity for making advances may be greatest.

National Research Council. 1996. *National Science Education Standards.* Washington, DC: National Academy Press.

Both perspectives result in limited and distorted ideas about the nature of science, not to mention the imagination and reasoning that so clearly characterize the human endeavor we call science.

Recent research suggests that use of the *National Science Education Standards* (National Research Council, 1996) as a guide for instructional practice and an understanding of the nature of science enhances the possibility of greater emphasis on evolution in science classes. Membership in a professional organization such as the National Science Teachers Association also enhances the possibility of greater emphasis on evolution in science classes (Weld and McNew, 1999). This evidence supports my recommendation to increase the emphasis on the nature of science in school science programs.

Science teachers who wish to respond to the challenge of teaching about the nature of science and evolution will find support in the *National Science Education Standards* (National Research Council, 1996, 200–204). Figure 1 presents an excerpt from the History and Nature of Science Standards for Grades 9–12. An excellent first step toward informing the public's discretion about science as a way of knowing would be to help students develop the understandings described in the figure. Clearly, using an inquiry-oriented approach would provide the foundation for developing these fundamental concepts and principles of science.

THE BEST REMEDY

Debating groups that advocate creationist positions for the school curriculum place science teachers in positions for which they are not prepared and are outside the realm of tasks the public expects of them. On the one hand, debating requires a huge expenditure of time and energy and results in little or no change among those advocating nonscientific positions. Also, in such debates there is every possibility of leaving creationists with more power and the public with greater confusion about science and science education.

On the other hand, teaching about the nature of science and incorporating inquiry in school science programs is directly aligned with the roles and responsibilities of science teaching. Teachers can address all students, and they can bring about significant change through developing public support for science. In this situation science teachers can win. Presenting an appropriate view of science and fulfilling the obligation to teach science benefits both students and society.

The best remedy for the continuous challenges from groups proposing that nonscientific ideas be incorporated in the science curriculum is a clear understanding of what science is and what it is not. Certainly, actual experiences with scientific inquiry in the classroom and lessons on the nature of science are appropriate for all students and achievable by all science teachers. If implemented in school science programs, this remedy avoids the slippery slope of arguments with creationists and builds a strong foundation for students and citizens who can exercise a wholesome discretion about science and science education.

REFERENCES

Ayala, F.J. 2000. Arguing for evolution. *The Science Teacher* 67(2): 30–32.

Baron, R.C. 1993. *Jefferson the Man: In His Own Words*. Golden, Colo. and Washington, D.C.: Fulcrum/Starwood Publishing.

Bruner, J. 1960. *The Process of Education*. Cambridge, Mass.: Harvard University Press.

Darwin, C. 1859. *On the Origin of Species by Means of Natural Selection or the Preservation of Favored Races in the Struggle for Life*. London: Murray.

Eldridge, N. 2000. *The Triumph of Evolution: And the Failure of Creationism*. New York: W.H. Freeman and Company.

Gould, S.J. 1999. *Rocks of Ages*. New York: Ballantine Publishing Group.

Lerner, L.S., and W.J. Bennetta. 1988. The treatment of theory in textbooks. *The Science Teacher* 55(3): 37–41.

Lewis, R. 1986. Teaching the theories of evolution. *The American Biology Teacher* 48(6): 344–347.

Mayr, E. 1991. *One Long Argument*. Cambridge, MA: Harvard University Press.

Moore, J.A. 1993. *Science as a Way of Knowing: The Foundations of Modern Biology*. Cambridge, MA: Harvard University Press.

Moore, J.A. 2000. Thought patterns in science and creationism. *The Science Teacher* 67(5): 37–40.

National Academy of Sciences. 1999. *Science and Creationism: A View from the National Academy of Sciences*. Washington, D.C.: National Academy Press.

National Academy of Sciences. 1998. *Teaching About Evolution and the Nature of Science*. Washington, D.C.: National Academy Press.

National Research Council. 1996. *National Science Education Standards*. Washington, D.C.: National Academy Press.

National Science Teachers Association. 1997. An NSTA Position Statement: The Teaching of Evolution. Arlington, VA: NSTA Webpage: *www.nsta.org*

National Science Teachers Association. 2000. *The Creation Controversy and the Science Classroom*. Arlington, VA: National Science Teachers Association.

Nord, W.A. 1999. Science, religion, and education. *Phi Delta Kappan* 80(1): 28–33.

Pennock, R.T. 1999. *Tower of Babel: The Evidence Against the New Creationism*. Cambridge, MA: Massachusetts Institute of Technology.

Weld, J., and J. McNew. 1999. Attitudes toward evolution. *The Science Teacher* 66(9): 27–31.

IT'S NOT JUST A THEORY

Why teachers need to address the nature of science and the "hidden" curriculum

DEWAYNE A. BACKHUS

"It's just a theory." This seemingly innocuous statement has a considerably different significance for me since the Kansas Board of Education action in August 1999, which used this phrase to diminish the role of evolution as a unifying theme in the Kansas Science Education Standards. The potential implications and consequences of statements like "It's just a theory" should not just be of interest to the science education community. It should be of concern for anyone in a position to influence the thinking of both students and the general population, either formally or informally.

Calling scientific knowledge "just a theory" greatly distorts the "nature of science," regardless of whether science is considered as a body of knowledge or a human endeavor. Theory is a "well substantiated explanation of some aspect of the natural world that can incorporate facts, laws, inferences, and tested hypothesis" (National Academy of Sciences, 1998). The "nature of science" envelops the qualities and processes that lead to these theories concerning our understanding of the sciences, including: assumptions, gathering evidence, making theoretical deductions, generalizing patterns and developing quantitative relationships, and formulating hypotheses.

The idea that evolution can be referred to as "just a theory" was instrumental in the thinking of lay members of the Kansas Board of Education when the August 1999 decision concerning the state's science education standards was promulgated. This misconception of what a theory is appears to

SECTION II
THE EDUCATIONAL PERSPECTIVE

> **Theory in the Classroom**
>
> Richard Duschl, author of *Restructuring Science Education: The Importance of Theories and Their Development*, asserts that instructional units in the sciences must transcend the teaching of facts and what some have called "final form" science. Instead, instruction should be congruent with the development of our conceptual understanding of science content—theories central to a subject—as organizing frameworks for instruction. Emphasis should be placed on the role of observations, the identification of patterns, the development of quantitative relationships when possible, and the formulation of theories that are often the sole focus of instruction.
>
> While this would necessitate a more arduous and time-consuming route to final form science, this epistemological approach could enrich understanding of, and appreciation for, a subject. Indeed, meeting standards and other teacher expectations is time-consuming. But perhaps the life science or biology teacher could organize at least one unit of instruction to provide greater conceptual understanding of the development of theory associated with biological evolution. Similarly, teachers of other science subjects could concentrate at least one unit of instruction on the development of theory appropriate to that subject; for example, the heliocentric theory, atomic theory, theory related to the behavior of light, or even theory associated with the life cycles of stars. The teaching of science would be enriched by providing foundations for our conceptual understanding based on evidence and reason and by presenting science in a fashion consistent with its nature as a human endeavor.

underlie thinking when a state's legislators, backed implicitly by the majority of its citizenry, tamper with textbooks and thereby influence the education of its youth. Ironically, the recent *National Science Education Standards* (National Research Council, 1996) and position statements of the National Science Teachers Association (2000) exhort a more central role for the nature of science and teaching science as inquiry than at any time since the post-Sputnik reawakening for science education in the late 1950s.

True, a reconstituted Kansas Board of Education reversed the previous board's action when it voted in February 2001 to return to the wording of state science education standards. This rewording was more consistent with the role of evolution as a foundation of the sciences expected to be taught and assessed in the state. But while the 1999 decision in Kansas became the proverbial "lightning rod" for activists intending to right a wrong, other states were harboring discourse and promulgating actions that paralleled the Kansas debacle. For example, the Alabama legislature recently mandated that all biology textbook adoptions continue to include a disclaimer questioning the validity of evolution as a conceptual theme for the biological sciences. (See *www.world-of-dawkins.com/Dawkins/Work/Articles/alabama/alabama.htm* for background on the Alabama circumstances.) However, Kansas and Alabama should not be considered alone. Various polls suggest that the public majority is uneasy with the precepts of evolution, and other evidence suggests that many science teachers have not reconciled their own thinking about the matter (Weld and McNew, 1999).

BEYOND COURSE CONTENT

Fortunately some actions are being taken to address these issues in the science curriculum. The Kansas science community published a monograph titled *A Kansan's Guide to Science: An Introduction to Evolution and the Nature of Science, Including Origins of the Universe and the Earth and the History of Life* (Kansas Geological Survey, 2000). *A Kansan's Guide to Science* parallels two publications of the National Academy of Sciences, *Teaching about Evolution and the Nature of Science* (1998) and *Science and Creationism* (1999). These monographs connect to the science curriculum: They reflect the intent to clarify science misconceptions from a content standpoint.

But curriculum embraces more than courses of study. Other opportunities exist for the classroom teacher at any level to address misconceptions related to the nature of science in general, and in particular to topics that are currently considered "controversial" or "sensitive." Teaching strategies that establish student understanding and insight into the nature of science need not be entirely pedantic (Duschl, 1990; Lederman, 1992; Lederman, Wade and Bell, 1998; and McComas, 1998) (See Theory in the Classroom, page 38). In addition, the standard idea of curriculum may be expanded upon to include a student's entire educational experience.

After reviewing several definitions of curriculum and their historic and social contexts, curriculum researcher P. W. Jackson concludes that curriculum might be inclusive of all "experiences" or "learning opportunities" that occur in schools "or under the guidance of teachers," and "not just those associated with the teaching of certain subjects..."(1992, 5). Others have reflected even more broadly concerning the curriculum and opportunities for influences on students' learning, whether positive or negative. Philosopher and educator John Dewey's (1902) conceptions of education would bring together the subject-centered basis of the traditional curriculum and the educational nature of a child's experiences in general. Curriculum pioneer, John Bobbit's (1918) general views of curriculum are inclusive of experiences from out-of-school and in-school settings. A more contemporary writer, Lawrence Cremin (1976), acknowledges the "educative" role of any social setting or context, such as the home, church or synagogue, library, museum, an organization such as the scouts, a day-care center, or the media, including the newspaper, radio, or television.

This research suggests a myriad of learning opportunities as a complement to the structured classroom activities of the formal curriculum. Jackson (1968) concedes to a "hidden curriculum" at the precollege level, and the Carnegie Foundation for the Advancement of Teaching recognizes curricula in three forms at the postsecondary level. One form is a "hidden curriculum, which consists of learning that is informally and sometimes inadvertently acquired by students in interactions with fellow students and faculty members" (1977, xiv). This type of learning supplements the formal curriculum of courses and the extracurriculum of less formal, college- or university-sponsored activities.

IMPLICATIONS FOR EDUCATORS

Because this hidden curriculum influences student learning, it is important that terms such as *theory*, and concepts such as the *nature of science*, are correctly used in various educational settings, not just in teaching science. It is misleading when a teacher or professor addresses a student with the question "What is your theory about this?" when the intent is to solicit an opinion. Yet these errors continue to surface. Recently, I became uneasy when a university chemist at a professional meeting opined—not theorized although he used the word *theory*—that his "theory" on this issue was.... I have also seen newspaper editorials on the "opinion" page describe a "conjecture" or "speculation" as a theory.

As science educators we are faced with some very subtle issues associated with our choice of words and their implications for desired or undesired learning outcomes. While the word *Theory*—let's think of

Topic: Newton's Laws
Go to: *www.scilinks.org*
Code: EIP03

it with a capital T—has a distinctive meaning for us in the sciences (National Academy of Sciences, 1998 and 1999), general use of the word *theory* has undermined our intent. Definitions provided in common dictionaries confound the issue that we face. In addition to an acceptable definition of *theory* as used in the sciences, one dictionary (*Merriam Webster's Collegiate Dictionary*, 1995) has variants which imply *theory* to mean speculation, a belief, conjecture, or that it is synonymous with hypothesis! Hence, if *theory* is considered to mean speculation, belief, hypothesis, or something akin to anyone's opinion, it is understandable that the statements it is "just a theory" or "only a theory" could be associated with the theory of evolution, and ultimately lead to it being discredited, as in

The Nature of Science and the Black Box

"Black box" activities are sometimes used to impart the spirit of investigation in the sciences.

The Introductory Physical Science text (Haber-Schaim et al., 1999) uses "black box" activities to illustrate the investigative nature of science. The student may work with a box that is about 12 cm x 12 cm x 4 cm. A typical box can include three metal rods, with two oriented parallel to each other in one of the long-box dimensions, and the other in nearly the same plane in the box oriented perpendicular to the two in the long-box dimension. Each rod is sufficiently long to protrude a couple of centimeters beyond each side of the box. The rods may be held in place with a rubber band strapped around the box from one end of each rod to the other. These rods can be labeled 1, 2, and 3 for the student's reference and purposes of communication. Three washers can be placed on the rods. An interesting arrangement is to place at least one washer at an intersection of two rods that are oriented perpendicular to each other.

After the box is closed and the students can only see the outside of the box, they are instructed to make observations (without, of course, opening the box) based on planned investigations. The students make accurate notes of what they observed when something specific was done to manipulate the box to lead to a conclusion. For example, the box might be rotated with one of the rods as the axis for rotation. Additional observations based on repeated systematic actions can lead to a student's "model" for the arrangement of rods and washers.

Then what shall be done with the box after the students have visualized its internal arrangement? In science, we frequently test a theory (or model) based on a prediction of a consequence of an action (a planned experiment, for example). If in the "black box" activity students predict that pulling a designated rod will lead to a particular consequence, actually pulling the rod will test the prediction. If all predictions lead to verifications, then the student has confidence in the model as visualized. If predictions are not confirmed, then students gain new information for changing the model. (It is difficult to deduce a washer at the intersection of two rods without actually doing predictions and the subsequent testing of the "black box" with its contents.) Of course in the process of predicting and testing, the original model is destroyed.

But this is consistent with the nature of science! To be most consistent with the nature of science and the scientific enterprise, the student investigator should not actually look into a box under any circumstances. Alas, it would not be consistent with the nature of science for a teacher to tell a student if a box was visualized "correctly"—that is, how the rods and washers were arranged before the student's investigation.

the case of the Kansas and Alabama legislature decisions.

What should be our response as scientists or science educators? We can engage students with inquiry-based activities explicitly intended to reinforce understanding concerning the nature of science; for example, the appropriate use of a "black box" (See the sidebar on page 40) or similar activities (Haber-Schaim et al., 1999; Boujaoude, 1995). However, as is the case with most effective, inquiry-based science instruction, we should not just tell students about these nuances related to the nature of science, but rather practice careful use of the language consistent with the nature of science and science as inquiry. Educators should not ask "What is your theory concerning. . . ?" but rather "What is your opinion concerning. . . ?" or "What are your thoughts concerning. . . ?"

If learning opportunities and influences abound—in the form of a truly "hidden" curriculum—we need to share these concerns with pastors, Sunday school teachers, librarians, museum assistants, zoo docents, scout leaders, newspaper editors, and radio and television commentators. The same vigilance should also pertain to the meaning of, and consequent use of, words such as *assumption, infer,* or *prove* when engaging in discourse concerning the nature of science. It's not just a theory.

REFERENCES

Bobbit, J.F. 1918. *The Curriculum.* New York: Arno Press.

Boujaoude, S. 1995. Demonstrating the nature of science. *The Science Teacher* 62(4): 46–49.

Carnegie Foundation for the Advancement of Teaching. 1977. *Missions of the College Curriculum.* San Francisco: Jossey-Bass Publishers.

Cremin, L.A. 1976. *Public Education.* New York: Basic Books.

Dewey, J. 1902. *The Child and the Curriculum.* Chicago: University of Chicago Press.

Duschl, R.A. 1990. *Restructuring Science Education: The Importance of Theories and Their Development.* New York: Teachers College Press, Columbia University.

Haber-Schaim, U., R. Cutting, H.G. Kirksey, and H. Pratt. 1999. *Introductory Physical Science.* Belmont, Mass.: Science Curriculum Inc.

Jackson, P.W. 1968. *Life in Classrooms.* New York: Holt, Rinehart, and Winston.

Jackson, P.W. 1992. Conceptions of curriculum and curriculum specialists. In *Handbook of Research on Curriculum,* Ed. P.W. Jackson. New York: Macmillan Publishing. pp. 3–40.

Kansas Geological Survey. 2000. *A Kansan's Guide to Science.* Lawrence, Kans.: University of Kansas.

Lederman, N.G. 1992. Students' and teachers' conceptions of the nature of science: A review of the research. *Journal of Research in Science Teaching* 29(4): 331–359.

Lederman, N.G., P. Wade, and R. Bell. 1998. Assessing understanding of the nature of science: A historical perspective. *Science & Education* 7(6): 595–615.

McComas, W.F. Ed. 1998. *The Nature of Science in Science Education: Rationales and Strategies.* Boston, Mass.: Kluwer Academic Publishers.

Mish, F.C. 1995. *Merriam-Webster's Collegiate Dictionary.* Springfield, Mass: Merriam-Webster Incorporated.

National Academy of Sciences. 1998. *Teaching About Evolution and the Nature of Science.* Washington, D.C.: National Academy Press.

National Academy of Sciences. 1999. *Science and Creationism.* Washington D.C.: National Academy Press.

National Research Council. 1996. *National Science Education Standards.* Washington, D.C.: National Academy Press.

National Science Teachers Association. 2000. *Position Statement on the Nature of Science.* Arlington, Va.: National Science Teachers Association.

Weld, J., and J.C. McNew. 1999. Attitudes toward evolution. *The Science Teacher* 66(9): 27–31.

EVOLUTION AND INTELLIGENT DESIGN

Understanding the issues surrounding evolution and intelligent design and dealing with the controversy

JOHN R. STAVER

In an ideal democracy all citizens would exhibit the capacity to agree to disagree and to tolerate and respect others' views. Though democratic, our society is far from ideal, and schools often become battlegrounds over contrasting viewpoints because public education simultaneously reflects society's broad landscape of values and transmits them as it prepares future citizens.

Recent school board actions in Ohio and Georgia concerning intelligent design (ID) theory in schools display all the earmarks of arguments over whose values should be taught in science class. In December 2002, following almost a year of national publicity and debate, the Ohio State Board of Education rejected attempts to include ID in Ohio's new science standards. In September 2002, the local board of education in Georgia's second largest school district, Cobb County, approved a policy affirming the discussion of contested viewpoints in academic disciplines, citing the origin of species as an example. In January 2003, after weeks of criticism from the scientific community, the board clarified its policy by stating that controversy over evolution is a social issue. As the 2003–2004 school year opened, attempts to include ID in school science were underway in Texas and New Mexico.

William Dembski, a leader in the ID movement, defines ID in three parts: "a scientific research program that investigates the effects of intelligent causes; an intellectual movement that challenges Darwinism and its naturalistic legacy; and a way of understanding divine action" (1999, 13). Michael Behe (1996), another proponent of ID, argues that many systems within living organisms are too complex to have developed by evolutionary processes; therefore, such systems must have been designed. Behe cites the bacterial flagellum, blood clotting pathway, and immune system as examples and coined the terms "irreducible complexity" and "irreducibly complex" to describe any "single system composed of several well-matched, interacting parts that contribute to the basic function, wherein the removal of any one of the parts causes the system to effectively cease functioning" (1996, 39).

Advocates of ID theory argue that evolution is a theory in crisis, ID is a legitimate scientific theory, and biology teachers should teach the controversy. Supporters of evolutionary theory testify that ID is a religious, not scientific, concept, and evolution is in no danger of bankruptcy, having survived 140 years of scientific scrutiny quite well. Evolution proponents further ask to examine ID advocates' scientific research reports on ID. This is indeed a marathon dispute; round one took place in a Dayton, Tennessee, courtroom in 1925.

This article discusses the process of generating new scientific knowledge and accepting it in our schools, explains why evolution is not in crisis, documents the lack of evidence in ID theory, and recommends methods to incorporate issues surrounding evolution in the classroom.

ESTABLISHING NEW KNOWLEDGE

The controversy between ID and evolutionary theory raises an important question: How is the subject matter of school science determined? The first step is the acceptance of new scientific knowledge by the relevant scientific community. Such knowledge may be in the form of a theory, law, principle, hypothesis, fact, or other aspects of the nature of science. When scientists amass extensive empirical evidence that conflicts with a prevailing theory, they begin to question that theory, and at some point they propose a new theory. Scientists in relevant fields empirically test the proposed theory's capacity to explain and predict. If this work eventually confirms that the new theory explains more, predicts better, and documents the limitations of its predecessor, then scientists incorporate the new theory by replacing or substantially modifying its predecessor.

One example often recounted in high school chemistry texts is the acceptance of the quantum mechanical model of the atom from its immediate predecessor, the Bohr model, and from the earlier models of Rutherford and Thomson. British scientist J. J. Thomson drew upon his knowledge of raisin pudding to conceive negatively charged electrons—analogous to raisins—as distributed throughout a consistent, positively charged mass of atom—analogous to pudding. Thomson's model predicted that alpha particles should be deflected when passed through a thin layer of atoms.

Testing Thomson's model, Ernest Rutherford passed alpha particles through thin gold foil and found that most alpha particles were undeflected. Rutherford concluded that an atom is primarily empty space with a small, dense, positively charged mass at its center. Rutherford was unable to reconcile the electron, but Niels Bohr proposed that an electron orbits the hydrogen nucleus in a circle at quantized energy levels. Bohr's model explained the spectrum of hydrogen but failed to explain the spectra of more complex systems. Louis deBroglie argued that electrons, like light, act as both particles and waves; the heavier the mass of a particle, the more negligible the wave character of its motion. DeBroglie's prediction that electron motion would demonstrate substantial wave character was confirmed by experiments

Topic: Rutherford Model of Atom
Go to: *www.scilinks.org*
Code: EIP04

observing the diffraction of electrons. The modern quantum mechanical model of the atom takes into account the substantial wave character of electron motion (Bodner and Pardue, 1989). This example of how scientific work modified or replaced older models and established the currently accepted model of electron motion in atoms illustrates how scientists empirically test and modify existing knowledge to produce new knowledge. Evolutionary theory is another example. Scientists also worked—and continue to work—to test, validate, and improve evolutionary theory as a tool to explain and predict biological phenomena.

Once established, new scientific understandings enter school science curricula after extensive discussions and consensus building among all stakeholder groups. Only after both steps are completed do new scientific theories, laws, principles, facts, and additional components of the nature of science reach schools. For example, the structure, function, and replication of DNA are routinely described in today's high school biology texts, but such descriptions are absent in the Gregor Mendel edition of *Modern Biology* (Moon et al., 1958). Their absence in 1958 is hardly surprising; Watson and Crick only proposed their model in 1953 and did not share a Nobel Prize with Wilkins until 1962.

Development of the *National Science Education Standards (NSES)* illustrates a more recent process of discussion and consensus (NRC, 1996). Twenty-two scientific and science education societies and over 18 000 individual scientists, science educators, teachers, school administrators, and parents reviewed and critiqued the efforts of the working groups of scientists, teachers, and educators who developed the Standards over a four-year process (NRC, 1997).

WHERE'S THE EVIDENCE?

ID dates back to Aristotle's final cause and the Stoics' inference of God's existence from biological complexity (Peterson, 2002). In 1802, the English philosopher and theologian William Paley published an argument by analogy: Just as a watch has a maker, so does the world have a maker, who is God (Gould, 1992). The modern concept of ID surfaced in 1991 when Phillip Johnson, now emeritus professor of law at the University of California at Berkeley, published *Darwin on Trial*.

Topic: Darwin and Natural Selection
Go to: *www.scilinks.org*
Code: EIP05

Twelve years have passed since Johnson subpoenaed evolution. If evolution is a theory in crisis, if scientists have proposed ID as a new theory, and if scientists in relevant fields are reporting research on the merits of ID theory, then we should find extensive empirical evidence in the form of published articles in refereed scientific research journals.

Leslie Lane, a biologist at the University of Nebraska–Lincoln, recently conducted an electronic search of the Science Citation Index over the past 12 years. Approximately 10 600 000 published articles were searched in 5300 journals. "Intelligent design" was only a keyword in 88 articles; 77 of these were in various fields of engineering, exactly where one should expect design to be a prominent concept. The remaining 11 articles included 8 that criticized the scientific foundation for ID; 3 of those articles appeared in nonresearch journals. "Specified complexity" and "irreducible complexity," two important concepts of ID theory, appeared in 0 and 6 articles, respectively.

On the contrary, approximately 115 000 articles used "evolution" as a keyword, primarily referring to biological evolution, and natural selection was a keyword phrase in 4100 articles (Lane, 2002a). In summary, there is no published evidence in refereed scientific literature surveyed that supports the hypothesis that the scientific community currently views evolution as a theory in crisis, with ID being studied as a potentially viable successor.

Lane conducted another electronic search of the Science Citation Index, Social Sciences Citation Index, and Arts and Humanities Citation Index over the last 12 years to examine ID advocates' admonition to teach the controversy over biological evolution within the scientific community. Approximately 13 800 000 articles were searched. Lane found only 22 articles with the keywords "creation" and "controversy," all focusing on the controversy between science and fundamentalist religion. Approximately 108 000 articles used the keyword "evolution," with perhaps one-fourth to one-third of these focusing on biological evolution (Lane, 2002b). This search demonstrates that the scientific community does not view evolution as a theory in crisis. Evidence does exist, however, for a controversy over evolution between religious fundamentalism and science.

NOT A SCIENTIFIC THEORY

ID is not a scientific theory for four reasons. First, ID's advocates do not set aside explanations that are beyond human reason, as the nature of science requires. Scientists refrain from considering God's actions in their work. Instead they use observation, experimentation, prevailing scientific theory, mathematics, logical arguments, strict empirical standards, heuristics, and healthy skepticism to produce knowledge. Scientific theories are therefore explanations about aspects of nature without reference to God. This means they are natural, and we call this context methodological naturalism.

Second, ID's advocates spend their time, resources, and energy criticizing and distorting the character of evolution in particular and science in general. One example is their claim that science practiced as methodological naturalism is atheistic. The methodological naturalism of scientific work and knowledge does not mean that scientific work and knowledge are atheistic. Let me demonstrate this point with two hypothetical science teachers, Sue and Joe. Do Sue and Joe consider God in deciding where to park their cars in the school parking lot or where to sit in a theatre? If not, Sue and Joe use methodological naturalism to make their decisions. Do Sue and Joe, then, reject God because they do not consider Him in their decisions? I think not. Scientific work and knowledge are properly silent about God because they do not consider Him. Science cannot logically conclude that God does or does not exist because science does not consider God in its work. Claims that evolution is atheistic or that science denies God are based on faulty logic, assume a particular philosophical view, or both.

The claim that evolution is a theory in crisis, perhaps even in bankruptcy, because it cannot answer certain questions is a classic example of a logical fallacy that philosophers call an appeal to ignorance. "When the premises of an argument state that nothing has been proved one way or the other about something, and the conclusion then makes a definite assertion about that thing, the argument commits an appeal to ignorance. The issue usually involves something that is incapable of being proved or something that has not yet been proved" (Hurley, 1997, 140). The further claim that ID is the only alternative to evolution involves the logical fallacy of the false dichotomy. This fallacy "is committed when one premise of an argument…presents two alternatives…as if no third alternative were possible" (Hurley 1997, 161). For instance, yet another alternative states that evolution is God's design.

Third, a useful scientific theory will be modified or replaced only when a new theory is proposed and documented through scientific work to explain and predict as much or more. (I discussed this issue above in terms of how new scientific understandings reach schools.) Until this happens, ID has no place in school science.

Fourth, the motivation, strategy, and behavior of ID advocates are not scientific. This point can be clarified with a metaphor. Standing in line is perhaps a universal experience. Meal lines, grocery cashier lines, driver's license lines, and auto license plate lines are all too familiar experiences that sometimes test our patience. What do folks who are waiting in line think when someone cuts in line? Their thoughts are probably neither positive nor charitable. ID theory and its advocates are attempting to cut in line. Biological evolution stood in line over 140 years to become the single accepted scientific theory for explaining the observed diversity and underlying unity of living organisms. Standing in line means that evolution has been thoroughly and repeatedly examined through accepted scientific work, which shows that evolution explains the observed diversity and underlying unity of living organisms very well, but not completely. Standing in line also means that scientists use evolution to make an extensive variety of accurate predictions. Such work has provided humans with antibiotics, vaccines, and better foods. ID advocates have not generated acceptable scientific work, nor have they set forth accurate predictions based on their work; their theoretical proposal is to an audience of citizens, not to the scientific community. A central strategy of ID advocates' appeal to the public is to charge that scientists are unfair, biased, close-minded, and protective of a bankrupt theory. The scientific community has responded, and continues to respond, by pointing out that the advocates of ID theory are attempting to cut in line by not conducting and reporting acceptable scientific work.

CONTROVERSY IN THE CLASSROOM

One question remains: How can high school science teachers apply these issues in the classroom? My short answer is to use the *NSES* as an instructional guide (NRC, 1996). Basing a brief expansion of my short response upon the work of Larry Scharmann (1993, 1994, 2001) and others, I maintain that teachers should first know and appreciate that students come to class with a great deal of prior knowledge. Time spent learning individual students' conceptions about evolutionary concepts, cultural values, and religious backgrounds is time well invested, as teachers can reduce perceived threats of evolutionary concepts and thereby help students become more receptive to understanding worldviews alternative to their own (Nelson, 1986; Scharmann, 1993, 1994; Meadows, Doester, and Jackson, 2000).

Teachers should also commit to teach for understanding, not belief. Teachers need to focus on helping students understand that scientific theories and scientific inquiry are specialized tools. Teachers should help students understand further that using the tools of science does not require the rejection of existing tools such as personal religious beliefs. The earlier series of questions regarding one's consideration of personal religious views in deciding where to park the car or sit in a theatre can resolve such conflicts. Simultaneously, teachers should employ proactive, student-centered instructional strategies, such as the Learning Cycle model for guided inquiry instruction (Scharmann, 2001). Such instructional methods engage students in scientific inquiry as they work to understand the facts, hypotheses, laws, and theories that such inquiry has produced. Guided inquiry methods are simultaneously more student-centered, standards-based, and constructivist-based.

Lastly, teachers should be prepared to capitalize on students' positive characteristics, their natural interests, and their demands for relevance. Teachers can accomplish this by employing guided inquiry methods and by emphasizing the power, utility, and relevance of evolutionary theory as an explanatory and predictive tool (Scharmann, 2001).

Evolution explains the extensive diversity of living organisms across the broad landscape of the life sciences and the unity that lies underneath such diversity. Predictions based on evolution have led to numerous advances that directly affect students' personal lives. Finally, teachers should perhaps point out that somewhere today is a high school student who may eventually develop a cure for HIV. These are the reasons to include evolution in the science curriculum, teach evolution in a Standards-based manner, and aim for all students to understand evolution.

REFERENCES

Behe, M.J. 1996. *Darwin's Black Box: The Biochemical Challenge to Evolution*. New York: Touchstone.

Bodner, G.M., and H.L. Pardue. 1989. *Chemistry: An Experimental Science*. New York: John Wiley and Sons.

Dembski, W.A. 1999. *Intelligent Design: The Bridge Between Science and Theology*. Downers Grove, Ill.: InterVarsity Press.

Gould, J.A. 1992. *Classical Philosophical Questions*. 9th ed. Upper Saddle River, N.J.: Prentice Hall.

Hurley, P.J. 1997. *A Concise Introduction to Logic*. 6th ed. Belmont, Calif.: Wadsworth Publishing Company.

Johnson, P.E. 1991. *Darwin on Trial*. Downers Grove, Ill.: InterVarsity Press.

Lane, L.C. 2002a. Intelligent design in the scientific literature. *www.geocities.com/lclane2/idlit.html*.

Lane, L.C. 2002b. Teaching the controversy. *www.geocities.com/lclane2/teachingthecontroversy.html*.

Meadows, L., E. Doester, and D.F. Jackson. 2000. Managing the conflict between evolution and religion. *American Biology Teacher* 62:102–107.

Moon, T.J., et al. 1958. *Modern Biology*. Ed. Gregor Mendel. New York: Henry Holt and Company.

National Research Council (NRC). 1996. *National Science Education Standards*. Washington, D.C.: National Academy Press.

National Research Council (NRC). 1997. *Introducing the National Science Education Standards*. Washington, D.C.: National Academy of Sciences.

Nelson, C.E. 1986. Creation, evolution, or both? A multiple model approach. In *Science and Creation: Geological, Theological, and Educational Perspectives,* ed. R.W. Hanson. New York: Macmillan Publishing Company.

Peterson, G.R. 2002. The intelligent-design movement: Science or ideology? *Zygon* 37(1): 7–23.

Scharmann, L.C. 1993. Teaching evolution: Designing successful instruction. *American Biology Teacher* 55(8): 481–486.

Scharmann, L.C. 1994. Teaching evolution: The influence of peer instructional modeling. *Journal of Science Teacher Education* 5(2): 66–76.

Scharmann, L.C. 2001. Preparing science teachers for anti-science actions: Avoiding red flags, using proactive strategies, and being prepared. Paper presented at the annual meeting of the Association for the Education of Teachers of Science, January, Costa Mesa, California.

ATTITUDES TOWARD EVOLUTION

Membership in professional organizations and standards use are associated with strong evolution teaching

JEFFREY WELD AND JILL C. McNEW

A recent science methods class at Oklahoma State University began quietly enough with a discussion of inquiry pedagogy, using an evolution activity as an example, but ended in vociferous debate about the merits of evolution theory itself. Wrangling over evolution and creationism is nothing new among high school students, but we were surprised to unearth polar discord among preservice science teachers—most of whom had amassed impressive records of course work in cell biology, ecology, vertebrate anatomy, evolution, and the rest of the standard fare for life science teacher preparation. A study was born at that point; we were determined not only to find out how much variation exists among teachers in the field regarding this controversial unifying concept but also to pinpoint the factors that shape teachers' inclinations to emphasize evolution in class.

Ironically, our findings would have immediate application for our neighbors to the north, where decisions to teach evolution are now left up to individual teachers and local school districts. On August 11, 1999, the Kansas Board of Education voted 6–4 to remove most references to evolution from the state standards, allowing local schools to choose whether or not to teach about biological diversity, the origin of the universe, and the development of the Earth (Blair, 1999).

SECTION III — THE SCIENCE TEACHER'S PERSPECTIVE

Science Teachers' Opinions on Evolution and Creationism

South Dakota
39% believe creationism should be taught in schools.

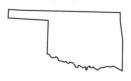

Oklahoma
33% place little or no emphasis on evolution.

Illinois
30% believe both creationism and evolution ought to be taught.

Kentucky, Indiana, Tennessee
23% place little or no emphasis on evolution.

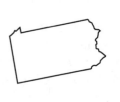

Pennsylvania
1/3 of science teachers do not think evolution is central to biology.

CURRENT VIEWS

Major policy documents are unequivocal in their treatment of evolution. Oklahoma's Science Teachers Association (OSTA) crafted a policy statement in 1981 that states, "Science teachers should not be required to teach, as science, ideas, models, and theories that are clearly extra-scientific. Any requirement for equal time for such hypotheses is not justifiable" (National Center for Science Education, 1995, 146). One year after the OSTA statement was released, the American Association for the Advancement of Science (AAAS) bolstered the case for teaching evolution in public schools in a resolution that demarks creationism from evolution, stating, "be it resolved that because 'Creation Science' has no scientific validity it should not be taught as science and further, that the AAAS views legislation requiring 'Creation Science' to be taught in public schools as a real and present threat to the integrity of education and the teaching of science" (National Center for Science Education, 1995, 25).

Amidst numerous court rulings between 1981 and 1995 that invalidated attempts to infuse the public school biology curriculum with theistic explanations for the origins and evolution of life, the National Science Teachers Association board of directors developed and approved a position statement on the teaching of evolution. This statement provides clear guidance on the place for evolution and creation (including the statement, "Science teachers should not advocate any religious view about creation, nor advocate the converse"), a background on what constitutes a scientific theory ("a theory is a set of universal statements that explains the natural world"), and the importance of evolution to understanding biology (citing the prominent role of evolution in the *National Science Education Standards*, AAAS's *Benchmarks for Science Literacy*, and NSTA's *Scope, Sequence, and Coordination Project*) (NSTA, 1997, 1–5).

The science education community consensus, both in Oklahoma and nationally, regarding the centrality of evolution to life science, has been amply demonstrated. A number of previous studies were conducted to determine how much allegiance to this consensus exists among science teachers. In Kentucky, Indiana, and Tennessee, 23.4 percent of science teachers report little or no emphasis on evolution in their courses (Ellis, 1983). Fifteen percent of biology teachers include a creationism component in their classes (Zimmerman, 1987). Thirty percent of Illinois science teachers believe that both evolution and creationism ought to be taught in science class (Nickels and Drummond, 1985). Thirty-nine percent of South Dakota biology teachers believe creationism ought to be taught in public schools (Tatina, 1989). One-third of Pennsylvania science teachers believe evolution is not central to the study of biology (Osif, 1997), and 23 percent of Louisiana biology teachers place little or no emphasis on evolution instruction (Aguillard, 1999).

Reasons for teachers' variable emphases on evolution have been the subject of several research studies. One study investigated the relationship between attitudes toward the theory of evolution and gender, duration of teaching career, urban versus rural teaching, degree attainment, and region of the country (Eve and Dunn, 1990). The researchers reported demographics, including educational background, to be "largely unrelated to the choice of an origins perspective" among teachers. An exception was a weak association between creationism inclinations and rural backgrounds (Eve and Dunn, 1990, 18).

In state science standards, there is a regional tendency to ignore, lightly address, or euphemistically refer to the theory of evolution in standards documents (Lerner, 1998). The state standards of Tennessee and Mississippi were found to ignore evolution. Arizona, Florida, and South Carolina addressed evolution "lightly." Georgia, Kentucky, and Alabama treat evolution with euphemisms, avoiding human evolution altogether (Lerner, 1998). Incidentally, Kansas received a high rating on the standards category, asking teachers to "eschew pseudo-science," reflective of a strong evolution component at the time of the study.

Factors associated with life science teachers' emphasis on evolution theory in Louisiana schools were examined, and an association was found between degree attainment and amount of biology coursework on the part of teachers, and the degree of emphasis they give evolution in their classes (Aguillard, 1999). Other studies suggest that additional coursework beyond the equivalent of an undergraduate major is insufficient for changing future science teachers' views on teaching evolution (Brumby, 1984; Eve and Dunn, 1990; Bishop and Anderson, 1990). This may be due to the ineffective manner in which complex concepts like evolution are taught (Brumby, 1984). Students hold fast to convictions based on moral, social, and religious commitments regardless of degree of understanding of such a controversial concept (Bishop and Anderson, 1990).

Our study extends the scope of this area of inquiry by examining the influence of professional memberships and use of the *National Science Education Standards* (National Research Council, 1996). We also revisited previously examined factors said to influence teachers' emphasis on evolution in their life science classes.

METHODOLOGY

Our study began with a questionnaire that included a number of features borrowed from similar statewide studies (Aguillard, 1999; Ellis, 1983) as well as verbatim phrases from policy documents such as the *NSTA Position Statement on the Teaching of Evolution* (NSTA, 1997) and *Science for All Americans* (Rutherford and Ahlgren, 1990). We included a number of items seeking demographic information (years of experience, gender, school size, college of preparation, and academic degree). We also sought data for cross-comparison purposes (including professional association memberships, background in the nature or philosophy of science, sources of perceived resistance to teaching evolution, familiarity with evolution and creationism institutes, preparedness to teach evolution, and use of the Standards). Finally, the bulk of the questionnaire consisted of ranking or strength of opinion items measuring teachers' emphasis on evolution or creationism in their courses. One open-ended query asked teachers to elaborate on their philosophies.

The instrument was piloted with teachers in the Stillwater, Oklahoma, region and with faculty in the Department of Zoology at Oklahoma State University. After making revisions based on their input, we mailed the questionnaire to a statewide sample of Oklahoma public secondary school life science teachers. Of a total pool of 840 secondary life science teachers in the state, we randomly selected 462 to receive our survey. We were able to ensure participants' anonymity by enclosing a return postcard along with a return envelope for the survey so that they could acknowledge completion of the survey separately from returning the survey itself. Ultimately, 224 completed surveys were returned after two separate mailings. In other words, 48.8 percent of our sample population returned the questionnaire, representing 26.7 percent of all life science teachers in the state.

OVERALL TRENDS

Most of the life science teachers in our sample population reported completing either coursework or independent study in the nature or philosophy of science (91 percent). About two-thirds of the teachers place moderate or strong emphasis on evolution theory, while nearly one-third place little, no, or counter emphasis on evolution theory in their biology courses (consistent with other studies nationally).

About one-fourth of Oklahoma public school life science teachers place moderate or strong emphasis on creationism. (Options for each response choice were defined as: Strong—I stress the theory of evolution throughout the course as the principle that ties together all aspects of biology; Moderate—I teach at least one unit about the theory of evolution and never avoid usage; Little—I rarely mention evolution except in response to student inquiry; No—I never initiate discussion and avoid use of the theory of evolution whenever possible; Counter—I provide evidence against the theory of evolution. A similar item measured emphasis on creationism by omitting the word "evolution" and inserting "creationism.")

Most life science teachers reported feeling well prepared to teach evolution (74 percent), and most agreed or strongly agreed with the National Association of Biology Teachers (NABT) policy statement that teaching biology requires classroom discussions and laboratory experiences on evolution (66 percent). Evolution is perceived by most to be a unifying theme in biology (57 percent), and a slim majority do not fear raising controversy by teaching evolution (58 percent). Teachers are nearly split over the existence of scientific evidence for creationism (48 percent agree or strongly agree that there is much scientific evidence for creationism), though most do not perceive creationism and evolution as equally viable scientific alternatives for explaining present life forms. The majority of public school life science teachers are unaware of the existence of a leading organization promoting creation science, the Institute for Creation Research. Among those who opted to respond to the item that asked whether they considered themselves constructivist teachers, most characterized themselves as constructivist science teachers, though a sizable proportion (N = 62) declined to respond to that item. Our instrument did not test reasons for nonresponses to this item, but several teachers commented in the margin that they were unfamiliar with the term *constructivism*.

Slightly more than half of respondents disagree or strongly disagree with the statement, "I use the *National Science Education Standards* in guiding instruction." Figure 1 represents the frequency of responses to the open-ended query among teachers who specified a particular stance on this issue.

COMPARISONS OF SUBPOPULATION TRENDS

We sought to address ambivalent research findings on relationships between attitudes toward the theory of evolution and gender, duration of teaching career, urban versus rural teaching, and degree attainment. Two-tailed *t*-tests of the difference between response frequency distributions (via *t*-test of equality of means) of these subpopulations in our study revealed no significant differences. In other words, we found no difference between the emphasis placed on evolution by males or females, new teachers versus veterans, rural versus urban and suburban teachers, or those with bachelor's versus those with master's degrees, when it comes to teaching evolution or creationism in public school biology.

Differences in response frequency distributions on a number of questionnaire items were found to exist between subpopulations, however. Secondary life sciences teachers who belonged to one or more professional associations answered several items significantly differently than did teachers who reported not belonging to any science teacher organization, by measure of *t*-tests for equality of means. Professional association members were more likely to agree that evolution is a unifying theme in biology class ($p = 0.002$; significance level > 0.05), and that the Standards guide their instruction ($p = 0.007$; significance level > 0.05). Those reporting no science education organization memberships were more likely to place emphasis on creationism instruction ($p = 0.002$; significance level > 0.05), and were more likely to agree that there is much scientific evidence for creationism ($p = 0.043$; significance level > 0.05).

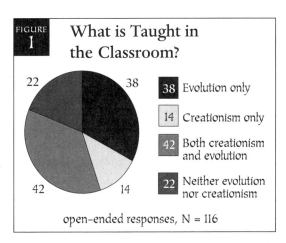

FIGURE 1. What is Taught in the Classroom?
- 38 Evolution only
- 14 Creationism only
- 42 Both creationism and evolution
- 22 Neither evolution nor creationism

open-ended responses, N = 116

The other subpopulation for which significant differences surfaced on questionnaire items was identified by splitting respondents into two groups—those who reported coursework or coursework and independent study in the nature or philosophy of science, and those who reported only independent study or no study whatsoever in the nature or philosophy of science. A strength of opinion item extracted from the NABT position statement on the teaching of evolution, "Teaching biology requires classroom discussions and laboratory experiences on evolution," was scored significantly differently between the two groups, as measured by *t*-test for equality of means (Minium, King, and Bear, 1993, 265). Teachers with no coursework in the nature or philosophy of science were more likely to disagree with the NABT statement ($p = 0.018$; significance level > 0.05).

Additionally, we examined possible relationships between demographic data and reported degree of emphasis to determine what factors may contribute to, or result from, particular views toward the teaching of evolution among life science teachers. Associations were found to exist between teachers' background in the nature or philosophy of science and the emphasis they place on the teaching of evolution. So, too, did associations surface between teachers' reported reliance on the Standards and the emphasis they place on evolution or creationism in the classroom. Also, an inverse correlation was found between level of study in the nature or philosophy of science and the perception of feeling less compelled to teach evolution because it is a theory and not a fact ($r = -0.138$; significance level > 0.05). Positive correlations exist between level of training in the nature or philosophy of science and the perception that biology requires discussions and laboratory experiences on evolution ($r = 0.228$; significance level > 0.05), and the attitude that evolution is a unifying central theme in biology ($r = 0.134$; significance level > 0.05).

The responses of teachers regarding their reliance on the Standards correlated negatively with the degree of emphasis placed on creationism in biology class ($r = -0.159$; significance level > 0.05). Positive correlations exist between reliance on the Standards in guiding instruction and the degree of emphasis placed on evolution in biology class ($r = 0.139$; significance level > 0.05), as well as the perception that biology requires discussions and laboratory experiences on evolution ($r = 0.172$; significance level > 0.05).

Cause and effect are difficult to pin down through an exploratory study such as this. Our results may imply that teachers who emphasize creationism are reluctant to join professional organizations, avoid courses in the nature of science, and shun the use of the Standards. Conversely, it could be that activities associated with organizational membership (journal reading, conferences, and so forth), plus participation in nature of science courses and guidance by the Standards influence the attitudes of science teachers.

As for implications for life science education, we could not possibly state the case more eloquently than did Theodosius Dobzhansky in his oft-quoted paper "Nothing Makes Sense Except in the Light of Evolution": "Seen in the light of evolution, biology is perhaps intellectually the most satisfying and inspiring science. Without that light it becomes a pile of sundry facts—some of them interesting or curious but making no meaningful picture as a whole" (Dobzhansky, 1973, 129).

Based on our findings that most biology teachers place moderate or strong emphasis on evolution, a majority of students of the life sciences is able to enjoy the intellectually satisfying and inspiring topic of evolution as either an integrated component or a stand-alone unit in biology class. Many of our students will therefore emerge from their public school experience more appreciative of the "meaningful picture as a whole," prepared as scientifically literate citizens to build upon evolution theory in formal or informal lifelong education. Yet misconceptions of this unifying theme in biology abound with many life science teachers, resulting in the misrepresentation of science and the customization of biology curriculum to individual tastes. Too many students of our public high schools are being denied the "light of evolution," thus perpetuating a 180-year-old misunderstanding that thwarts intellectual potential.

The findings of our study contribute to the rich body of literature in this area in several ways. First, we found that emphasis placed on teaching evolution or creationism does not appear to be tied to demographic data such as teachers' gender, the size of their communities, the duration of their careers, degree attainment, or the type of college from which they graduated. Second, the fact that most of our teachers feel well prepared to teach evolution suggests more content coursework than that typically encountered en route to licensure and certification may have little influence on tendencies to teach evolution or creationism. Echoing a research citation discussed earlier, this finding could likely be a function of ineffective college course pedagogy. Third, the factors most strongly associated with teachers' philosophies toward emphasizing evolution in their classrooms are

- coursework and independent study in the nature or philosophy of science;
- membership in professional science teacher organizations; and
- use of the Standards in guiding their practice.

When the data are isolated so that we examine only the responses of teachers who report coursework in the nature or philosophy of science plus membership in one or more associations plus use of the Standards in guiding their instruction (N = 35), three-fourths place moderate or strong emphasis on evolution in their classes (as opposed to 66 percent overall).

The limitations of this study preclude making prescriptions toward professional membership or guidance by the Standards as solutions to the problem of creationism in public school science class. But we are able to predict with reasonable assurance that biology teachers who fit this profile are more likely to accurately represent the scientific explanation for the evolution of life on Earth. The stage is set for further analysis of the influence of peer associations and the other factors influencing adoption of Standards-based practice, as they impact biology teachers' personal philosophies toward evolution theory.

ACKNOWLEDGMENT

This project was funded through an Oklahoma State University College of Education Dean's Research Support Grant.

REFERENCES

Aguillard, D.A. 1999. Evolution education in Louisiana public schools: A decade following *Edwards v. Aguillard*. *The American Biology Teacher* 61(3):182–188.

Bishop, B. A., and C. W. Anderson. 1990. Student conceptions of natural selection and its role in evolution. *Journal of Research in Science Teaching* 27(5):415–427.

Blair, J. 1999. Kansas evolution controversy gives rise to national debate. *Education Week* 19(1):1, 24–25.

Brumby, M. N. 1984. Misconceptions about the concept of natural selection by medical biology students. *Science Education* 68(4): 493–503.

Dobzhansky, T. 1973. Nothing makes sense except in the light of evolution. *The American Biology Teacher* 35:129.

Ellis, W. E. 1983. Biology teachers and border state beliefs. *Society* 9(6): 26–30.

Eve, R. A., and D. Dunn. 1990. Psychic powers, astrology, and creationism in the classroom? *The American Biology Teacher* 52(1): 10–21.

Lerner, L. S. 1998. State science standards: An appraisal of science standards in 36 states. Report of the Thomas B. Fordham Foundation. March. Available online at *www.edexcellence.net/library/nbpts.html*.

Minium, E.W., B. M. King, and G. Bear. 1993. *Statistical Reasoning in Psychology and Education*. New York: John Wiley and Sons.

National Center for Science Education. 1995. *Voices for Evolution*. Berkeley, CA: National Center for Science Education.

National Research Council. 1996. *National Science Education Standards*. National Academy Press.

National Science Teachers Association. 1997. *Position Statement on the Teaching of Evolution*. Available online at *www.nsta.org*.

Nickels, M., and B. Drummond. 1985. Creation/evolution: Results of a survey conducted at the 1983 ITA convention. *ISTA Spectrum* 11–15.

Osif, B. A. 1997. Evolution and religious beliefs: A survey of Pennsylvania high school teachers. *The American Biology Teacher* 59(9):552–556.

Rutherford, F. J., and A. Ahlgren. 1990. *Science for All Americans*. New York: Oxford University Press.

Tatina, R. 1989. South Dakota high school biology teachers and the teaching of evolution and creationism. *The American Biology Teacher* 51:275–280.

Zimmerman, M. 1987. The evolution-creation controversy: Opinions of Ohio high school biology teachers. *Ohio Journal of Science* 7:115–121.

THE SCIENCE TEACHER'S PERSPECTIVE

SECTION III

INVESTIGATING ISLAND EVOLUTION

A Galapagos-based lesson using the 5E instructional model

ANTHONY V. DEFINA

iking slowly over the rocky terrain, I tune into the whistling and honking sounds around me, being careful not to step on the radiating blue feet that are literally in my path. I am surrounded by a colony of blue-footed boobies intent on the business of courtship dances and displays. Just when I think it can't get any better, a male red-flanked marine iguana gently steps over my boot as he makes his way toward the water's edge to feed on algae.

This was my experience when I traveled to the Galapagos Islands for the first time and was awestruck by the uniqueness of these "enchanted isles" and the natural lessons they present. A biology teacher and naturalist at heart, I have an affinity for wild places and for birding, photographing plants and animals, and learning about species and their natural history. In my classroom, nature serves as a tool to connect students to relevant biological concepts and processes through anecdotes, discussion, multimedia, and studying alluring species. Many endemic species of the Galapagos Islands are captivating and useful for a biology teacher planning an extended lesson on evolution.

Because the *National Science Education Standards* (National Research Council, 1996) promote teaching science through inquiry, I wanted to use my experience in the Galapagos to design a lesson that allows students to recognize and work through problems, gather information, think critically and analytically, and draw conclusions based on evidence and supported by explanations. Students may have preconceived ideas and sometimes misconceptions about evolution from their previous experiences, so a lesson plan must reflect the contemporary constructionist view of learning wherein students redefine, reorganize, expand, and change their initial concepts and views through classroom activities and experiences while interacting with other students and their environment. To satisfy these concerns, I followed the 5E instructional model (Trowbridge et al., 2000). This model describes a sequence of phases—engagement, exploration, explanation, elaboration, and evaluation—that align with both inquiry processes and constructivism.

ENGAGING ISLANDS

During the engagement phase, the challenge is to gain students' attention, spark their inquisitiveness, and have them pose initial questions. Showing an enticing video clip, passing around photographs, projecting a slide show of an attractive species such as the Galapagos hawk (*Buteo galapagoensis*), or describing a spectacle such as that of the male great frigatebird (*Fregata minor ridgwayi*) in his courtship ritual work for me as a way to accomplish this task. I then have students pose their own questions about the features, location, wildlife, and scientific significance of the islands to ascertain their prior knowledge. This discussion generates some useful Galapagos facts, and presenting a brief history and description of "Darwin's Islands" further serves to establish a base of knowledge.

Once students have been engaged, they can be given some geographical information about the island. These oceanic islands form an archipelago that consists of 13 relatively large islands, each with an area greater than 10 km², 6 smaller islands, and more than 40 named islets. Formed from undersea volcanoes between three to five million years ago, the islands today straddle the equator some 960 km from Ecuador on the South American mainland (Jackson, 2001). Because they are sitting out in the middle of nowhere, so to speak, the question arises as to how plants and animals colonized and inhabited these islands, especially in view of the geologic fact that these volcanic islands were barren when they first appeared above the sea's surface.

At this point, student exchanges and discussion focus attention on dispersal mechanisms and island colonization. By the end of this introduction, the class acquires insights into possible dispersal mechanisms to the Galapagos from potential mainland areas; for example, by sea for penguins and marine mammals; through floating vegetation rafts for insects and reptiles; by air for seabirds; and by wind for plants, lichens, and fungi. In addition, weak flyers such as some land birds and bats could be blown out to sea and carried to the islands by strong winds. The stage is then set for students to delve further into Galapagos natural history as they tackle evolution-based questions during the exploration phase of the model.

CLASSROOM EXPLORATION

To initiate this stage, students participate in role-playing. They form groups and become explorers and naturalists, imagining that they are part of Charles Darwin's famous voyage to the Galapagos. Their adventure revolves around resources at their disposal that include map diagrams of the Galapagos; photographs of endemic species and volcanic landscapes; Galapagos natural history books and biology texts; pertinent articles from available magazines such as *Natural History, National Geographic, Audubon, Wildlife Conservation,* and *International Wildlife*; Galapagos filmstrips and video programs from sources such as National Geographic ("Dragons of the Galapagos," 2001) and the PBS nature series; and a listing of Galapagos Website links. In addition, students have a series of species notes—observations made of some of the native species while

Topic: Galapagos Islands
Go to: *www.scilinks.org*
Code: EIP06

exploring the islands with Darwin (Figure 1). Armed with these resource materials and species notes, students act as investigators to research, reflect, and interact in collaborative groups to develop scenarios, hypotheses, and plausible mechanisms that can explain the occurrence, appearance, and development of endemic Galapagos species.

During this constructive and active exploration phase, the teacher can monitor and facilitate group interactions, guide individuals to resources, and encourage collaboration without providing direct instruction. Students discover things for themselves as they dissect the species observations and review resources to look for common themes and patterns. It is also a time to listen as students discuss their ideas and perceptions about the mechanisms they think are at work. This not only provides a preview to their thoughts and explanations for how evolution works, but it also identifies possible misconceptions that must be addressed in the next phase: explanation.

STUDENTS EXPLAIN EVOLUTION

Following the exploration, groups present their explanations and scenarios in their own words to the class, and their interpretations serve as a basis for establishing new learning. For example, as students research and discuss notes on the endemic marine iguana, penguin, boobies, cormorant, and sea lion, they typically identify common factors and forces at work over time. These include having traits that allow for survival in new habitats, being able to survive in isolated environments, and having the reproductive ability to establish numbers of individuals. Their presen-

Species Notes of Galapagos Wildlife with Suggestions for Student Exploration.

• The marine iguana (*Amblyrhynchus cristatus*) occurs only in the Galapagos. It grazes on algae and is the only sea-going lizard in the world. Students can debate how the first colonizing iguanas survived and how such a marine lizard could develop over time in the Galapagos.

• The odd Galapagos penguin (*Spheniscus mendiculus*) occurs only in the Galapagos and is related to the Humboldt and Magellan penguins found in southern South America. Questions arise related to its survival along the equator and what accounts for the differences that separate it from its relatives.

• Three seabird species, the blue-footed booby (*Sula nebouxi excisa*), the red-footed booby (*Sula sula websteri*), and masked booby (*Sula dactylatra granti*) occur elsewhere in the world, but the populations found in the Galapagos form a taxonomically distinct subgroup. Student groups can exchange views on what makes these varieties different from their relatives in other areas of the world, and how the distinctions occur over time in the Galapagos.

• Compared to other cormorant species in the world, the flightless cormorant (*Nannopterum harrisi*) is the only one to have lost its ability to fly. Students theorize about colonizing cormorants and how and why such birds changed over time in the Galapagos.

• The delightful Galapagos sea lion (*Zalophus californianus wollebacki*) exists only in the Galapagos, but it is classified as a subgroup of the California sea lion. Group exchanges focus on how and why this species became established.

tations and commentaries provide a launching pad to present scientific explanations and terminology that focus on the following concepts (Starr and Taggart, 2001; Purves et al., 1998):

- Species and subspecies;
- Species diversity;
- The meaning and genetic basis of variations and adaptations;
- The process of Darwinian natural selection that shapes species through descent with modification;
- The significance of the environment in establishing the conditions for survival and filtering traits that are adaptive;
- The effects of isolation, especially on geographically isolated oceanic islands.

ELABORATION

Application and expansion of these concepts and ideas characterize the elaboration phase. During this segment, students have an opportunity to correctly employ technical vocabulary and apply scientific reasoning to new but similar situations. Students revisit their Galapagos Islands exploration and see that additional species notes merit their attention and require their investigation to identify a common framework and potential evolutionary mechanisms. Before long, the class again refers to resources and engages in meaningful group conversations as they respectfully argue their own views and relate what they think they know to these new scenarios. Students explain the development and emergence of one

---

FIGURE 2 **Species Notes Used during the Elaboration Phase.**

• Thirteen endemic, distinct species of Darwin's finches (*Geospiza spp.*, *Camarhynchus spp.*, *Platyspiza crassirostris*, and *Certhidea olivacea*) inhabit the Galapagos and are thought to be descended from a single South American blue-black grassquit ancestor. They vary in feeding habits and beak structure.

• Of the four endemic species of Galapagos mockingbirds (*Nesomimus spp.*), one of each populates the islands of Espanola, San Cristobal, and Champion Island off of Floreana, respectively; and the fourth species is distributed as six subspecies on most of the remaining islands.

• The seven endemic species of lava lizards (*Tropidurus spp.*) most likely evolved from a single ancestral species. These species differ remarkably in the precise pattern of a behavior displayed by both sexes that mimics a human exercise-like "push-up"; this pattern is performed more actively and frequently by males during aggression or courtship. Of six species, one of each resides on the islands of Floreana, San Cristobal, Pinta, Marchena, Espanola, and Santa Fe; the seventh species occurs on 10 other islands of the Galapagos.

• There are 11 different subspecies or races of the endemic Galapagos giant tortoise (*Geochelone elephantopus*). Five of these varieties exist, respectively, on five distinct volcanoes of Isabella Island, while the other six races individually occupy six separate islands. They vary in the shape of their carapace or upper shell.

• Of the two endemic species of Galapagos land iguanas, *Conolophus pallidus* occurs only on Santa Fe, and *Conolophus subcristatus* occurs on five other islands of the archipelago. They feed mainly on prickly pear cactus.

or more of the endemic Galapagos species, subspecies, and varieties listed in Figure 2 (Jackson, 2001; Castro and Phillips, 1996).

After examining these species, representatives from each student group use maps of the archipelago to present their interpretations and conclusions. For example, most groups propose that the endemic Galapagos finches, mockingbirds, lava lizards, giant tortoises, and land iguanas separated from their respective common ancestor species to form new species or varieties within specific environmental conditions. Students should employ correct use of concepts such as adaptation, natural selection, and isolation as they propose scenarios.

I use these concepts and students' misguided notions to create a teaching base to help students learn about speciation and adaptive radiation. For example, their comments and proposals open a pathway for instruction not only on the reproductive and genetic isolating mechanisms of allopatric or geographic speciation, but also on comparable mechanisms for sympatric and parapatric speciation. In addition, student interpretations provide avenues to introduce scientific explanations for the influences of genetic drift and founder effects on small colonizing populations, and the concept of ecological speciation based on small-scale accommodations to everyday survival (Starr and Taggart, 2001; Purves et al., 1998). Opportunities also arise to introduce environmental issues that focus on the human introduction of alien species such as feral goats, pigs, dogs, and cats into the Galapagos, and, consequently, their destructive impacts on native species.

STUDENT EVALUATION

The final phase of the sequence calls for evaluation. During each phase of the lesson, I assess the knowledge gained by students by listening to the adequacy of their explanations and viewpoints and looking for indications that they have changed their thinking. In addition, other formative assessment tools, such as short open-ended

FIGURE 3. A general rubric used for the scoring of evolution essays.

Content/concepts (Maximum 5 points each)
species and subspecies
variations and genetic sources
adaptations
natural selection
environment and conditions for survival
speciation processes
geographic isolation factor
reproductive isolation factor
genetic isolation factor
founder effect/genetic drift and populations
common ancestor
adaptive radiation process
colonization of oceanic islands

Presentation/format
organization of topics and subtopics
sentence structure/grammar
spelling and punctuation
use of observations to support statements
use of additional evidence to support statements
overall presentation and writing format
overall use of scientific terms and scientific thinking

Total

> Many endemic species of the Galapagos Islands are useful for a biology teacher planning an extended lesson on evolution.

questions provide feedback on comprehension and application of scientific terms and evolutionary concepts.

However, for a summative assessment, I prefer to use a two-part essay. This assignment consists of both an in-class examination and a take-home research component. Students work individually in class to construct a well-written essay that challenges them to apply learned vocabulary and scientific concepts and to analyze species observations. For example, they may be asked to discuss the evolutionary processes and factors that account for the distribution and variations of Galapagos tortoises. The second part of the essay requires a research assignment to be completed out of class and presents a similar challenge. For example, students must investigate and then discuss the evolutionary processes that explain the distribution and variations of fruit flies or honey creeper birds on the Hawaiian Islands. Moreover, they must proceed to draw conclusions, make comparisons, and indicate contrasts about processes observed in the Galapagos. The essays are scored according to a 100-point rubric (Figure 3). As part of an overall evaluation, it is also important for students to self-evaluate their learning through reflection and perhaps journal writing.

Biology has its historic roots in whole plants and animals interacting in their respective ecosystems. As more professional biologists specialize in cellular and molecular areas, fewer consider themselves to be naturalists. Although biotechnology and bioinformatics are important, they should not be emphasized at the expense of organismic biology. Biology teachers are in a unique position to bridge the gap between molecular and organismic biology. Using a "laboratory of evolution" such as the Galapagos can help construct that bridge for students.

REFERENCES

Castro, I., and A. Phillips. 1996. *A Guide to the Birds of the Galapagos Islands*. Princeton, N.J.: Princeton University Press.

"Dragons of the Galapagos." 2001. National Geographic Television, Washington, D.C.

Flannery, M. 2001. Where is biology? *The American Biology Teacher* 63(6): 442–447.

Jackson, M.H. 2001. *Galapagos: A Natural History*. Calgary, Canada: The University of Calgary Press.

National Research Council. 1996. *The National Science Education Standards*. Washington, D.C.: National Academy Press.

Purves, W., G. Orians, H. Heller, and D. Sadava. 1998. *Life: The Science of Biology*. Sunderland, Mass.: Sinauer Associates.

Starr, C., and R. Taggart. 2001. *Biology: The Unity and Diversity of Life*. Pacific Grove, Calif.: Brooks/Cole.

Trowbridge, L., R. Bybee, and J. Powell. 2000. *Teaching Secondary School Science: Strategies for Developing Scientific Literacy*. N.J.: Prentice Hall.

THE SCIENCE TEACHER'S PERSPECTIVE

SECTION III

SEARCHING FOR THE PERFECT LESSON

Teaching evolution to a diverse biology class

SUSAN STONE PLATI

It's one minute to class time. A group of high school students sits at a table examining data from the previous day's lab. James, a tall boy with a ponytail, gets up to retrieve additional data from one of the classroom's computers. While James records the data, Karen arrives with a late note. Several more students follow quickly on her heels; their geometry test ran five minutes beyond the bell. Meanwhile, Hua Lin records information about the feeding behavior of goldfish in the small aquarium. In the midst of it all, Maria approaches the teacher with a special request: Could she have a letter of recommendation for a science program for gifted students? Of course the answer is yes, but the teacher needs to begin class: "Now that you're all here, I'd like to explain what we're going to do today..."

Welcome to my classroom. It's probably similar to thousands of biology classrooms across the country. My class is a very diverse group of sophomore biology students. Each of them has unique needs, aspirations, and abilities, and to teach these students is a joyful experience. Engaging such a heterogeneous group in active learning and cooperative activities is a rewarding—albeit difficult—job. I find myself refining my craft every year. As I reflect on pedagogy that works, I also reexamine how the course content is sequenced. Often a slight change in the sequence can foster greater understanding in a larger number of students.

Content sequence matters a great deal in the subject of evolution, which forms the basis of biology. For more than 30 years, I have taught evolution and natural selection at the beginning of the course. All subsequent topics can be explained in light of evolution. Students learn about Darwin and evolutionary evidence in September; the diversity of habitats and life forms in October; and—after studying reproduction—Mendelian genetics, energy, the mechanisms of population genetics, and evolution in the spring. I also realized early on in my teaching career that students learn best in an activity-based program that involves much data gathering and analysis. As I select books and materials for class, I tend to choose those that are based in evolution and contain engaging activities. Or if such activities are lacking, I look for a flexible program that allows me to integrate various activities that I have designed or learned from colleagues.

ENGAGING ALL STUDENTS

My best teaching happens when students are actively involved in their own learning. In a heterogeneous class, where students have different experiences and abilities, many activities can involve students at various levels. For example, my students learn about fossils, one of the pieces of evidence for evolution, by searching through containers of sediments containing samples collected at Calvert Cliffs, Maryland. This provides tactile learners with a true field experience. By looking at photographs of the cliffs, students also can visualize the rock columns that produced these fossils.

Other activities done in class include using Darwin's theories. An important part of Darwin's evolutionary idea states that more individuals are born than can survive, that there are variations among the individuals in a population, and that surviving individuals possess favorable variations that they pass on to their offspring. The survival of the organisms best suited to a particular environment is called natural selection. To illustrate the various aspects of evolution by natural selection, students study population growth patterns by growing fruit flies in culture bottles. They learn about variation by measuring differences in lengths of peanut cotyledons, height of humans, thumb size, and pumpkin mass. They study selection of resistant bacteria by growing bacteria in the presence of various antibiotics. They watch film clips of birds feeding on moths and selecting the ones that are most visible. They participate in feeding simulations of various types.

During all of these activities, students work in groups composed of classmates with various learning styles and are encouraged to discuss the data and reflect on the meaning of the data with their group. They also debate and produce newspapers written from the perspective of a British reporter in the late 1800s who has just been introduced to the *Origin of Species*. Here is where the creative and artistic students often come to the fore.

Because these activities interest students and engage them in authentic science activities, I should be satisfied, but I'm not. Like the baseball pitcher in pursuit of the perfect game or the inventor trying to build a better mousetrap, I'm a teacher looking for a better lesson and a timely curriculum. Imagine my excitement whenever I discover a resource that helps me do both.

A few years ago, I participated in the Woodrow Wilson National Fellowship Foundation Summer Institute in Evolutionary Biology for high school teachers. For one month, 50 teachers from all over the United States met at Princeton University to study evolution. We heard speakers, carried out laboratory investigations, discussed curriculum, and wrote lessons to take back to our classrooms to help teach evolution. We learned how to integrate lessons in evolutionary biology into all aspects of the biology curriculum.

SCILINKS
THE WORLD'S A CLICK AWAY
Topic: Explore Mendelian Genetics
Go to: *www.scilinks.org*
Code: EIP07

For example, our night search for spiders led to information about the evolution of reproductive behavior—perfect for a unit on reproduction. Our day search for dandelions and other plants on the Princeton campus led to a DNA analysis of these plants to determine the genetic similarity and evolutionary relationship of these organisms. Some members of the group developed various activities using this molecular evidence to help students understand evolution as well as the structure and function of DNA. This enhanced knowledge and holistic approach to studying evolution has caused me to ask questions of my students—questions that help them realize that understanding evolution is fundamental to understanding biology.

My school has also supported its science teachers in the effort to build a better curriculum. Three years ago, the biology teachers in our school adopted a curriculum that met all of our ideal criteria—engaging, inquiry-based activities that allowed for flexibility and provided challenging content based on evolution (Biological Sciences Curriculum Study, 1997). We have used this curriculum for the past two years with considerable success. Students work in groups, share information, design experiments, and engage in a wide range of activities from traditional labs to computer-based simulations to videodisc sequences of human and nonhuman primates. We allow students who wish to pursue this course at the honor level to sign a contract that reflects the added responsibilities of an honor student. The honor and regular college preparatory students are together in the same class. They learn from each other, help each other, and push each other to succeed.

> Like the inventor trying to build a better mousetrap, I'm a teacher looking for a better lesson.

MULTIMEDIA FOR MANY LEARNERS

More recently, I was introduced to the Evolution Project, which includes a seven-part, eight hour public television series co-produced by WGBH Boston and Clear Blue Sky Productions that debuted on PBS in September (*Evolution*, 2001). When I first heard of the project, I thought it would be a great video resource for my classes. But when I later became involved as a teacher-advisor to the team developing some of the teacher and student support materials, I realized that the Evolution Project is far more than just a television series. It involves the presentation of materials to enhance the teaching and learning of evolution in a manner that is completely compatible with the heterogeneous classroom.

For example, a comprehensive eight-session online course for high school science teachers draws on all aspects of the Evolution Project to help deepen our own understanding of evolutionary concepts and to address obstacles to teaching evolution sometimes faced in the heterogeneous classroom. The activities in both the teacher's guide and online student lessons speak to various learning styles—from a wonderful activity that asks students to look at the "evidence" of their own lives as part of a study on fossils, to an online lesson that sends students to an interactive Website to explore and evaluate the different extinction hypotheses surrounding the demise of the dinosaurs. Most students in my classroom are comfortable using the computer for various learning tasks. They enjoy interactive sites and Internet research.

Although I have found many useful resources to improve my teaching, I'm still searching for the perfect lesson and trying to build a better curriculum that will work for all my students, addressing their different interests, aptitudes, and learning styles. That's why teachers never get bored. The search for perfection is an endless, and endlessly rewarding, pursuit.

REFERENCES

Biological Sciences Curriculum Study. 1997. *Biology: A Human Approach*. Dubuque, Iowa: Kendall/Hunt Publishing.

Evolution. 2001. Boston, Mass.: WGBH and Clear Blue Sky Productions; *www.pbs.org/evolution*.

Woodrow Wilson National Fellowship Foundation. 1995. *Evolution: A Context for Biology*. Princeton, N.J.: Woodrow Wilson National Fellowship Foundation. Available online at *www.woodrow.org/teachers/biology*.

COMPARING COMMON ORIGINS

Using biotechnology to teach evolution

JOHN McLAUGHLIN AND GEORGE GLASSON

For years, educators have struggled to teach evolution in an inquiry-based fashion. Typically, students learn about evolution from textbooks, films, or models without examining direct evidence in nature. According to the *National Science Education Standards* (National Research Council, 1996), the teaching of biological evolution as inquiry presents a pedagogical challenge for science teachers. Fortunately, with the emerging field of biotechnology, science educators now have a more conceptual approach to and laboratory tools for teaching biological evolution in an empirical fashion. Students can use protein electrophoresis and access genetic and taxonomic databases to investigate molecular variation and examine evidence for biological evolution and common origins.

LOOKING FOR LINKS

Biotechnology, the use of living materials to solve problems or make useful products (Kreuzer and Massey, 1996), allows scientists, teachers, and students to study organisms at the molecular level and to use comparative analyses of DNA to learn how organisms might be related evolutionarily. For example, scientists would expect two different organisms that have similar DNA molecules to have diverged recently in time, whereas two organisms that do not have similar molecules most likely diverged much earlier on the evolutionary calendar.

SECTION III

THE SCIENCE TEACHER'S PERSPECTIVE

SciLinks
THE WORLD'S A CLICK AWAY

Topic: Comparing Common Origins
Go to: *www.scilinks.org*
Code: EIP08

While DNA is responsible for coding for specific proteins, to some degree, all living things share similar proteins. When a cell divides, random errors in DNA nucleotide sequences occur. These errors may lead to the emergence of different proteins in different organisms. Therefore, the more differences in protein biology two organisms share, the more time has elapsed between the two organisms' divergence. We would also expect two organisms that share a high level of protein similarity to have similar anatomical features and be closely linked phylogenetically.

Using a process called electrophoresis, scientists can easily isolate molecular fragments, the most common being nucleic acids (DNA and RNA) and proteins, from living material. In protein electrophoresis, fragments of proteins are separated on an acrylamide gel by the size of the molecules, and they appear on the gel as bands (Figure 1). The larger the bands, the more of a certain size of protein is present in the sample. Two samples that appear to have similar bands likely share similar proteins and probably are closely linked evolutionarily. By separating protein molecules, scientists can determine which organisms share the most identical material and theorize about lineages of different species.

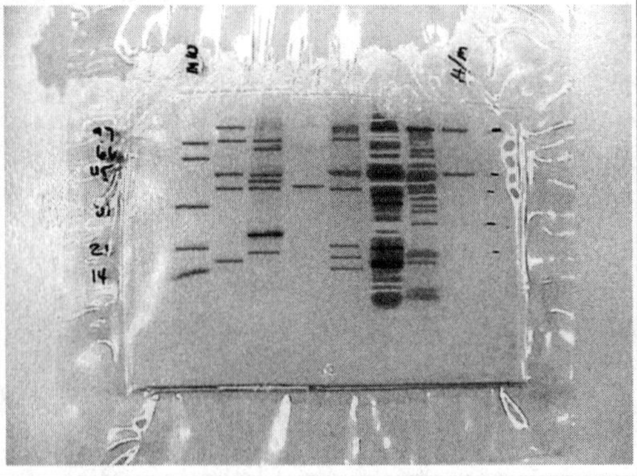

Students can also learn about phylogenetic relationships of organisms through the emerging field of bioinformatics, in which scientists use genetic databases to study genetics and molecular biology. Internet-based genetic databases of DNA nucleotide base sequences of various organisms allow students to compare evolutionary relationships of different species.

PROTEIN ELECTROPHORESIS LAB

Students can take advantage of biotechnology by doing a protein electrophoresis lab that teaches about evolution (DeCourcy, 1999). This interdisciplinary approach is appropriate for general biology classes or an AP Biology class. Using a protein electrophoresis lab combined with information from genetic databases allows teachers to teach evolutionary concepts in an inquiry-based fashion (National Academy of Sciences, 1998). In this lab, students learn how to isolate and compare different proteins from the muscle tissue of marine vertebrates and invertebrates to determine phylogenetic relationships. By separating and analyzing protein bands on an acrylamide gel, students can look for similarities and differences among organisms, and, in the process, construct a phylogenetic tree (Figure 2).

Students can also access databases of DNA nucleotide sequences of the different organisms to make inferences about changing relationships among organisms over time. Using samples of muscle tissue from seafood readily available at the local grocery gives excellent results. Students can use mollusks (clams and oysters), crustaceans (crabs and lobsters), cartilaginous fish (sharks), and bony fish (flounder and trout).

Many local colleges and universities have protein electrophoresis equipment that high school teachers can borrow. Because a class will work with about six samples, only one apparatus is necessary to perform this experi-

FIGURE 1 Protein Electophoresis Gel with Bands.

ment. Many tools, such as sample loading guides and disposable micropipettes, make the experiment easier than most teachers may think. In Virginia, for example, the Fralin Biotechnology Center (*www.biotech.vt.edu*) on the campus of Virginia Tech has an outreach department where kits for protein electrophoresis, DNA electrophoresis, plant tissue culturing, and chromatography are available at no charge to Virginia teachers who attend a half-day workshop for each topic. The kits contain all solutions, materials, and equipment for as many as eight groups of students.

Biotech Websites can also be extremely helpful to teachers. Sites such as the one produced by the National Center for Biotechnology Information (NCBI), on the Web at *www.ncbi.nlm.nih.gov*, are tremendous resources for teachers and students. Such sites have helpful teaching tips, educational resources for teachers, and entire tutorials through which students can enhance their skills in a variety of biotech concepts.

Phase One: Engagement
Our first step in this activity is to engage students by asking them to make hypotheses based on

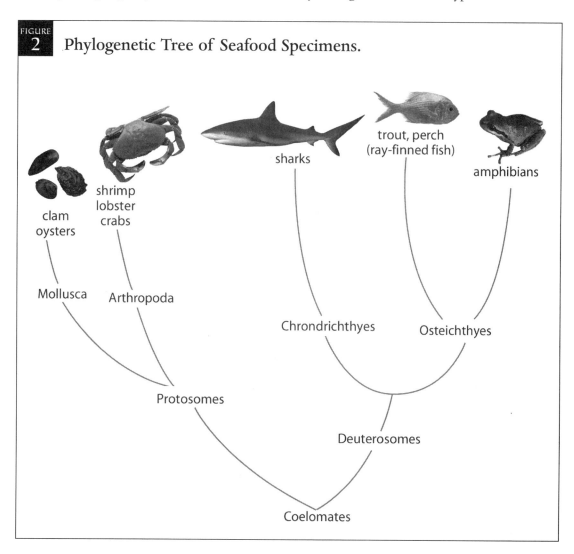

FIGURE 2. Phylogenetic Tree of Seafood Specimens.

common characteristics of the sample organisms. Teachers should ask students to choose a type of sea creature to analyze, and they may want to ask students to bring some type of seafood to class. Students should do library, textbook, or Internet research to learn more about the animal's physical traits and be able to share with the class its most notable common characteristics. Teachers can actively engage students in scientific inquiry by having them construct rudimentary phylogenetic trees using anatomical evidence that they can gather from research. Based on these trees, students can formulate a hypothesis for the lab. For instance, they might formulate a hypothesis that says, "Sample one and sample two are placed side-by-side on the phylogenetic tree because of like characteristics." As the lesson progresses, the teacher may want to have students begin to write discriminatory explanation on lines connecting organisms.

Phase Two: Exploration

The second stage of this activity is that of exploration. Students perform a sodium dodecyl sulfate (SDS) gel protein electrophoresis. Protein electrophoresis is a standard laboratory procedure in which charged protein molecules are separated while traveling through an electrical field. In this experiment, a polyacrylamide gel acts to separate the molecules of protein by their

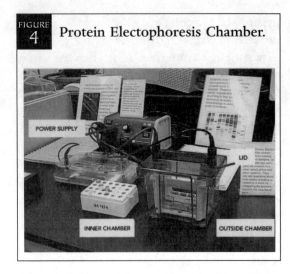

FIGURE 4 — Protein Electophoresis Chamber.

shape and charge. The smaller the molecule, the faster it will travel through the gel matrix. Because proteins are typically charged, an SDS buffer is used in this type of electrophoresis, allowing for only mass to be the separating element. SDS allows proteins to separate as if they had identical charges.

The gel is actually made of two parts; one is a stacking gel, which is on top, and the other is a separating gel, which is on the bottom (Figure 3). The stacking gel has a lower acrylamide concentration, different pH, and different molarity of buffer. This is called a discontinuous buffer system, and it allows each sample of protein to enter the separating gel at the same time. Rather than have students pour their own gels, prepared polyacrylamide gels are used in a vertical electrophoresis apparatus (Figure 4). The gel comes prepackaged between two plastic plates. Within the apparatus, the upper chamber with the negative electrode and the bottom chamber with the positive electrode are connected only by the acrylamide gel. The gel is clamped into position, and a running buffer is added to both the upper and lower chamber.

The samples are prepared by grinding down small samples and adding buffer to break apart proteins. The wells within the gel are filled with protein sample and molecular marker sample.

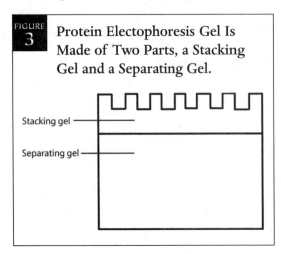

FIGURE 3 — Protein Electrophoresis Gel Is Made of Two Parts, a Stacking Gel and a Separating Gel.

The gel box is connected to a power source and is run at 150 V for about 45 minutes.

Once the gel has run, it can be dried and analyzed. Students can easily see which bands each organism has in common and more importantly, which bands may be dissimilar. By analyzing these bands, students can make inferences about an organism's protein biology and evolutionary development. A complete lab protocol can be found on Virginia Tech's Fralin Biotechnology Center's Website *(www.biotech.vt.edu/outreach/outreach.html)*.

Phase Three: Explanation

The third phase of this activity entails explanation—students analyze the protein electrophoresis data. Using the molecular weight marker as a guide, students draw lines from each band to the nearest molecular weight band and identify some of the proteins by their molecular weight (Figure 5). The predominant protein in each of the samples will be muscle protein. Because each sample is so similar, it is important to reinforce the fact it is impossible to identify a protein by its molecular weight alone. Therefore, further research is necessary to substantiate an evolutionary comparison of the samples.

Students should determine which samples have bands in common. Even if a student cannot identify the proteins by the marker, they should still record band similarities among samples. Then, knowing which samples share like proteins, students can return to their original phylogenetic trees based on physical traits to see if the protein data matches the characteristic data and if their hypotheses were correct.

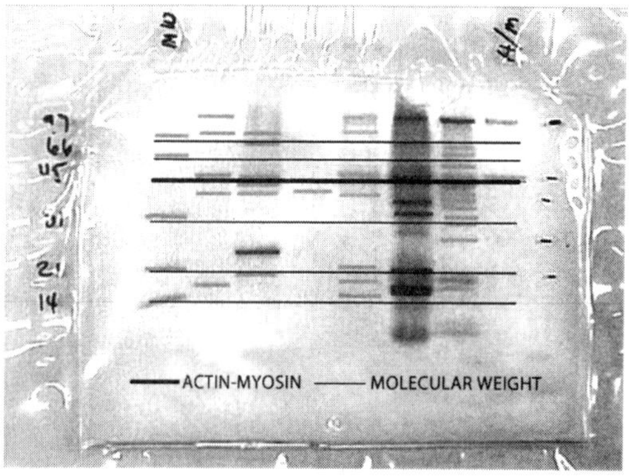

Through concept mapping or open-ended questioning, either in class or in the form of a lab worksheet, teachers can direct students to making the connections among DNA, proteins, traits, and changes over time. Students should be reminded that finding organisms that show similar protein banding patterns in an electrophoresis gel does not clearly define lineage. It simply identifies a potential relationship between the two organisms. Using proper questioning, teachers should engage students in open-ended inquiry into the type of relationship. One simple concept students should understand is that the more varied the samples are, the more differentiation will be seen in the protein bands on the gel. To visualize this relationship, students can do a general comparison of protein bands among samples on the gel to construct a phylogenetic tree. This phylogenetic tree (based on protein band similarities) can be compared to students' original trees constructed according to physical traits.

FIGURE 5 Protein Gel marked to Show Differences between Different Species.

Phase Four: Extension

The fourth phase of this activity is an extension. At this point, students should compare their phylogenetic trees to cladograms developed by scientists using protein similarities, fossil records, homologous structures, and DNA evidence. A good source of such cladograms is the Tree of Life Website *(phylogeny.arizona.edu/tree/)*. To use this Website, students should identify the scientific genus and species name for each organism represented by the class samples and then print out a copy of the appropriate tree that best matches the samples used. Using the phylogenetic tree they constructed based on the gel separation data, students can examine the scientists' cladograms and compare those to their own trees.

Another good resource at this point is the NCBI Website *(www.ncbi.nlm.nih.gov/)*. Students can enter the genus name of the organism represented by a sample in the submission blank. Students will then come to a page in which they can select a partial nucleotide sequence of an organism. Students can compare and contrast differences in DNA sequences and make inferences about the evolutionary history of the organism.

Phase Five: Evaluation

The final stage is the evaluation stage. Within this learning cycle model, evaluation information may be collected throughout the laboratory investigation. Teachers can identify the scientific inquiry competencies (concepts and processes) addressed in the lab and the evidence that students have met the particular competencies (Figure 6). Student proficiency with these competencies provides evidence of an empirical basis for students' understandings of evolution. The more data students can retrieve and process, the more clearly they will be able to make the connections between the biochemistry of proteins and the process of evolution.

The field of biotechnology creates exciting new pathways for students to engage in scientific inquiry. By using protein electrophoresis to compare proteins among organisms, students can examine empirical evidence to infer evolutionary relationships. With the emerging field of bioinformatics, students can examine molecular data collected by scientists to learn more common origins of species.

REFERENCES

DeCourcy, K. 1999. *Protein Electrophoresis Kit: Information Manual*. Blacksburg, VA: Fralin Biotechnology Center.

Kreuzer, H., and A. Massey. 1996. *Recombinant DNA and Biotechnology*. Washington D.C.: ASM Press.

National Academy of Sciences. 1998. *Teaching about Evolution and the Nature of Science*. Washington, D.C.: National Academy Press.

National Research Council. 1996. *National Science Education Standards*. Washington, D.C.: National Academy Press.

FIGURE 6: Scientific Concepts, Processes, and Student Evidence

Scientific concepts of processes	Evidence
Students will make predictions based on prior knowledge.	Students will construct hypothetical phylogenetic trees.
Students will be able to collect and record scientific data.	Students will have dried protein gels.
Students will make observations.	Observations notes should be recorded during lab.
Students will develop understanding of how proteins control visible traits.	Students will note similar traits of samples.
Students will develop an understanding of how proteins may change through time.	Students will note that evolutionarily older samples have more different protein banding patterns.

THE SCIENCE TEACHER'S PERSPECTIVE

SECTION III

SYMBIOSIS: AN EVOLUTIONARY INNOVATOR

Blurring the concept of individuality, symbiosis tangles the phylogenetic trees

EMILY CASE

hen I plan a lesson or evaluate a curriculum, foremost in my mind are the *National Science Education Standards (NSES)* and the Massachusetts Curriculum Frameworks. My course syllabus, textbook choice, assessment methods, and instructional style are all influenced by these documents. My job is to teach the concepts and skills scientists and educators at the national and state levels have agreed are most important, and I take that responsibility seriously.

Some aspects of the national and state science curriculum guides, however, make me uncomfortable. Science is dynamic and skeptical; new discoveries and interpretations emerge continuously and must withstand rigorous tests before they replace old ideas. Science is also rooted in our cultural, social, and political context; the questions we ask, the research we fund, the explanations we believe are all shaped by our time and place. Because of all these factors, the explanation with the most supporting evidence is not always the one currently accepted by the scientific community. This phenomenon, of course, trickles down to science education standards. Every year, an example of this problem stops me in my lesson-planning tracks: the role of symbiosis in evolution.

DEFINING SYMBIOSIS

If the subject of symbiosis even appears in a textbook or state curriculum, it is grouped with ecology. Short-term costs and benefits of symbiosis are usually described—parasitism, mutualism, and commensalism. But symbiosis is best

SECTION III   THE SCIENCE TEACHER'S PERSPECTIVE

FIGURE 1 The British Soldier Lichen, *Cladonia cristatella* (a) and Its Constituents, the Fungus of the Same Name (b) and Green Alga *Trebouxia* (c).

Lichens are symbiotic organisms that consist of a fungus and either a green alga or a cyanobacterium. Note how the morphology of the lichen differs from the appearance of the both the fungus and the alga grown individually. Photos courtesy of Lynn Margulis (a), Vernon Ahmadjian (b), and Jim Deacon (c).

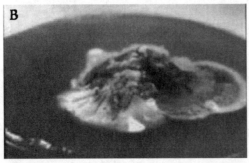

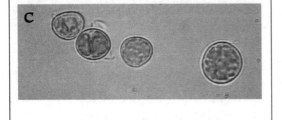

defined as a physical association between, or among, organisms of different species that persists for most of their life history. By this definition, pollination is not an example of symbiosis because the bee and the flower are not in prolonged physical contact. Lichens, however, are symbiotic because the alga, or cyanobacterium, and the fungus live together as one (Figure 1). Short-term costs and benefits are nothing to the evolutionary implications of symbiosis, and fascinating things happen when organisms spend a lot of time in or on one another.

SYMBIOSIS AND EVOLUTION

Mention of the evolutionary implications of symbiosis might bring to mind the symbiotic origin of chloroplasts and mitochondria. Most high school biology textbooks have an article (in a box in the margin of a page in the cell chapter) that describes the theory that mitochondria and chloroplasts were once free-living bacteria that were engulfed by larger cells. The organelles retain their own DNA and have multiple membranes attesting to their former independence and subsequent ingestion. This phenomenon—when symbiosis results in the formation of a new kind of being—is called symbiogenesis. But why is this symbiogenesis story in a box? Why is it in the cell chapter instead of the evolution chapter? And why is symbiogenesis absent from state and national standards?

Here's why: most scientists ignore the evolutionary implications of symbiosis beyond these events that produced mitochondria and chloroplasts. But symbiosis, it turns out, has everything to do with evolution by natural selection. Darwin's theory of natural selection requires populations to be genetically diverse; variation is the raw material for evolutionary change. The *NSES* (and almost any biology textbook) describes two sources of

Topic: Symbiosis
Go to: *www.scilinks.org*
Code: EIP09

74   NATIONAL SCIENCE TEACHERS ASSOCIATION

genetic variability. The relevant standard for students in grades 9–12 is as follows:

> Species evolve over time. Evolution is the consequence of the interactions of (1) the potential for a species to increase its numbers, (2) the genetic variability of offspring due to mutation and recombination of genes, (3) a finite supply of the resources required for life, and (4) the ensuing selection by the environment of those offspring better able to survive and leave offspring (NRC, 1996, 185).

Who could find fault with this? This standard is straightforward Darwinian selection, right? Except for one crucial detail. Darwin didn't (and couldn't) identify the source of genetic variability. That detail has been filled in by the "modern synthesis" of Darwinian selection with Mendelian genetics. Most biologists, biology teachers, and standard-writers are satisfied with the explanation that mutation and recombination of genes are the only sources of inherited variation. But in the case of the symbiotic origins of mitochondria and chloroplasts, symbiosis, not incremental mutation, provided the genetic variability. And talk about genetic variability—this was no mere base-pair change, but the introduction and integration of an entire genome. The result, ultimately, was the eukaryotic cells of protists, animals, plants, and fungi.

SHAPING THE GENETIC LANDSCAPE

The symbioses that produced mitochondria and chloroplasts are unquestionably significant. They are not unique. Symbiosis always has been and continues to be a major force shaping the genetic landscape.

For one thing, symbiosis is ubiquitous in nature. A multitude of symbioses can be observed in a teaspoon full of pond water, a walk in the woods, or a peek at the contents of human intestines. Further, in symbiosis, the whole is very different, if not greater, from the sum of its parts.

Lichens, for example, consist of over 30 000 different organisms with properties one never could have predicted from observations of either the fungus or its photosynthetic partner alone. Such unique properties result from the interaction of genomes; the potential for variability here is far greater than from mutations in a single gene or from recombining genes from individuals of the same species.

Topic: Symbiosis of Fungi
Go to: *www.scilinks.org*
Code: EIP010

What began as a simple physical association eventually became interaction on the genomic level in some symbioses. The symbiosis between the nitrogen-fixing bacteria *Rhizobium* and legumes, such as beans or clover, involves physical association (the bacteria take up residence inside the root cells for the growing season), exchange of metabolic products (the plant receives nitrogen-rich organic compounds from the bacteria,

FIGURE 2 *Trifolium pratense*

Legumes like this common red clover, *Trifolium pratense*, have symbiotic bacteria (*Rhizobium*) in their roots that convert atmospheric nitrogen into organic compounds necessary for plant growth. Photo courtesy of the author.

the bacteria take sugars from the plant), and an amazing genetic collaboration (Figure 2).

Without collaboration on the genetic level, the *Rhizobium*–legume symbiosis simply wouldn't work. The bacterial enzymes that fix nitrogen are destroyed by oxygen. Photosynthesis in plants, of course, produces oxygen. Nitrogen fixation is protected from oxygen by a molecule called leghemoglobin, similar to the hemoglobin found in human blood. The bacteria live in round swellings on the roots of the legume. The inside of these swollen root nodules often has a pinkish tint, due to the presence of leghemoglobin (Figure 3). The leghemoglobin binds and removes oxygen from the root nodule, providing a safe haven for the nitrogen-fixing enzymes. Here's the trick: the *heme* molecule is synthesized by the bacteria, the *globin* by the plant.

Symbiosis blurs the concept of individuality and tangles the phylogenetic trees. Is that lichen one organism or two? Is the lichen a fungus or an alga? Symbiosis becomes symbiogenesis when a physical association becomes heritable, passed on through generations like any other genetic trait. A new kind of life emerges. The scientific literature shows that symbiosis has been implicated in past and present speciation, and may very well account for the sudden appearances of new forms in the fossil record, a pattern known as punctuated equilibrium.

Darwin's missing detail—the source of genetic variability—is no small matter. Natural selection is a subtractive force. The potential of populations to increase in size despite their finite resources results in "the ensuing selection by the environment of those offspring better able to survive and leave offspring" (NRC, 1996, 185). If natural selection subtracts from existing genetic variability, what adds to it? What are the creative forces behind evolutionary change, those that produce new structures, body plans, capabilities, and species?

Anything that alters genes can be considered a creative evolutionary force, and thus mutations and recombination certainly fit the bill. But consider symbiosis as a source of genetic variability. The integration of entire genomes, with their thousands of tried-and-true, functional genes, results in the sudden appearance of novel biological forms. Contrast this with the gradual accumulation of random mutations. Mutations generally affect a single gene and are almost always deleterious. The potential for change is low, the potential for change that increases an organism's survival and reproduction lower still.

The enormous potential of symbiosis as a creative evolutionary force is evidenced in the plethora of existing associations such as the lichen, morphologically and functionally different from its constituent organisms, or the collaborative synthesis of leghemoglobin by bacterial and plant genomes. Symbiosis created eukaryotic cells and is implicated in field and laboratory studies of speciation. Symbiosis plays a starring role in the evolutionary drama, but has only a walk-on part in our science education curricula. So what should teachers do?

FIGURE 3 **Clover Root Nodules.**

These clover root nodules house nitrogen-fixing *Rhizobium* and pinkish leghemoglobin, a molecule that ferries oxygen away from the root nodule and the oxygen-sensitive enzymes responsible for nitrogen fixation. The leghemoglobin is jointly synthesized by the plant and the bacterial genomes. Each nodule is about 1–3 mm in length. Photo courtesy of Jim Deacon.

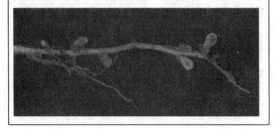

ROOM FOR DISSENT

First, of course, I have some suggestions for the policymakers who tell teachers what to teach. Obviously, we must add symbiosis as a source of genetic variation to our standards. But the problem is deeper than just this one conceptual flaw in the curricula. There is no single body of knowledge that everyone, even (or perhaps especially) scientists and science educators, will agree is free from conceptual flaws. There needs to be room in a state-mandated curriculum for dissent. I believe this is possible.

Policymakers must make sure the standards are truly a "core" curriculum, designed to provide the heart of but not the whole course of study. A good core curriculum ensures that all students learn certain, fundamental concepts but leaves room for teachers to supplement with lessons of local interest or personal expertise, delve into some topics in depth, and pursue the teachable moment. Assessments designed to test content knowledge must allow students to express what they think and know, not merely pick their brains for facts and terminology.

The *NSES,* and most states' curricula, already acknowledge the dynamic and skeptical nature of science. For example, the History and Nature of Science Content Standards (NRC, 1996, 200) require students to understand how scientific ideas become established and how science is done. Teachers too often focus only on subject area standards, which provide ample material for syllabi. But it is possible to do both. The symbioses described in this article—just the tip of the iceberg—are rich in biological concepts: metabolism, gene expression, and evolution. A student who understands symbiosis as an ecological and evolutionary phenomenon, and questions why it is misrepresented in his or her textbook, gains a depth of knowledge about both the content and process of science that would be missed by a student more traditionally schooled.

The learning standards of my state's mandated, core curriculum form the basis of my course syllabus. However, my prerogative, and my responsibility, is to critique these standards. My students will know most biologists believe that mutation and recombination are the main sources of genetic variability upon which natural selection acts. But they'll also know about the enormous power of symbiosis as an evolutionary innovator. And in learning to question the discrepancy between beliefs and evidence, students will learn a great deal about the history and nature of science.

ACKNOWLEDGMENTS

Feedback, support, and figures for this article came from Lynn Margulis and members of the Margulis Lab at the University of Massachusetts Amherst, Jim Deacon of the University of Edinburgh, and Lisa Poirier.

REFERENCES

Brodo, I.M., S.D. Sharnoff, and S. Sharnoff. 2001. *Lichens of North America.* New Haven: Yale University Press.

Deacon, J. The Microbial World. *helios.bto.ed.ac.uk/bto/microbes/microbes.htm.*

Margulis, L., and D. Sagan. 2002. *Acquiring Genomes: A Theory of the Origins of Species.* New York: Basic Books.

National Research Council (NRC). 1996. *National Science Education Standards.* Washington, D.C.: National Academy Press.

Wolfe, D.W. 2001. Out of thin air. *Natural History* 110(7): 44–53.

A TEACHING GUIDE TO EVOLUTION

Discovering evolution through molecular evidence

THOMAS G. GREGG, GARY R. JANSSEN, AND J. K. BHATTACHARJEE

Evolution is considered by virtually all biologists to be the central unifying principle of biology, yet its fundamental concepts are not widely understood or widely disseminated. Teaching evolution—defined as descent with modification from a common ancestor as a result of natural selection acting on genetic variation—has traditionally been a challenge for most high school biology teachers and has become even more controversial and difficult recently. This is largely due to pressure against teaching evolution by many school boards, school administrators, and parents, and by a highly organized well-funded campaign by creationist groups. The challenge has been intensified by the periodic and recent discoveries of hominid fossils directly linking human origins to the evolutionary process (Asfaw et al., 2002).

For many students their high school education represents their only formal introduction to scientific theories and explanations of phenomena from the biological world. Those continuing their science education, however, must have an understanding of the basic principles of evolution. Based on 10 years of teaching both a summer DNA workshop and an evolution workshop for teachers, it is clear to us that many biology teachers have unfortunately not been well prepared in either the theory or the evidence for evolution. This is because, as the teachers themselves assert, most college and university biology courses do not deal specifically with evolution. This is especially true of the molecular evidence, which constitutes some of the

SECTION III

THE SCIENCE TEACHER'S PERSPECTIVE

best evidence for evolution. The purpose of this article is to provide a short summary of some of the most compelling molecular evidence for evolution in hopes that it might be useful to biology teachers at all levels.

MOLECULAR EVIDENCE OF EVOLUTIONARY RELATIONSHIPS

Universality of DNA Structure

The molecular evidence in support of the evolutionary relationship among all living organisms is based upon clearly demonstrated, well understood, universally accepted scientific discoveries that can be verified again and again by students and teachers alike. Geneticists and biochemists have learned that deoxyribonucleic acid (DNA) is the genetic material, and genes consist of nucleotide sequences in the double helix of DNA that specify the sequence of amino acids found in proteins. With the exception of some ribonucleic acid (RNA) viruses, DNA is the genetic material of all living organisms, including archaea, bacteria, protists, fungi, plants, and animals, including humans.

The primary structure of DNA is composed of deoxyribonucleotides linked to each other by phosphodiester linkage. The secondary structure of DNA consists of two antiparallel polynucleotide strands, which are hydrogen bonded to each other by the complementary base pairing between adenine (A) of one strand and thymine (T) of the other polynucleotide strand; likewise, a guanine (G) of one strand pairs with a cytosine (C) of the other polynucleotide strand. The general process of DNA synthesis, and the complex structure of DNA consisting of exactly the same four nucleotides, are conserved in all organisms and have not changed in more than three billion years of life on Earth.

Topic: Observing Traits Molecularly
Go to: www.scilinks.org
Code: EIP11

Similarities among Diverse Organisms

Results from comparing the gene sequences for the small subunit ribosomal RNA, 16S in procaryotes or 18S in eucaryotes, from a wide variety of organisms shows that they all fall into three distinct clusters of similarity that define the three biological domains, Archaea, Bacteria, and Eucarya (Figure 1) (Woese, 1998).

Despite this distinct grouping of all organisms into three domains they nonetheless share a number of features that indicate their common ancestry.

The complete sequencing of the human genome is now regarded as a major milestone in the history of scientific accomplishment (Lander et al., 2001; Venter et al., 2001). This feat and the technical advances that accompanied it have made it possible to sequence a number of other species as well and to make sequence comparisons among them. This accomplishment has opened a whole new field of scientific investigation called "comparative genomics." Comparisons of more than 60 species representing all three domains have revealed a high degree of sequence similarity and genetic relatedness among these widely diverse organisms.

One result of particular interest is that eucaryotic nuclear genes are similar to those of the Archaea, whereas the mitochondrial genes of eucaryotes are similar to those of Bacteria. This is one line of evidence supporting the recent idea that the original relationship that gave rise to the first eucaryotes involved an organism in the Archaea engulfing a bacterium (Horiike et al., 2001; Martin and Muller, 1998), although the still current paradigm is that mitochondria arose as endosymbionts in eucaryotes. Eukaryotic genes consist of numerous coding regions—exons—that are separated by noncoding regions—introns. The entire gene is transcribed into a pre-messenger RNA, from which the intron sequences are removed, joining the exons together into a single messenger RNA molecule.

Proteins consist of interrelated but somewhat independent functional domains. Research from the genome project shows that exons in genes correspond to functional domains in proteins. In human genes 90 percent of the exons are homologous to exons found in *Drosophila* (fruit flies) and *Caenorhabditis* (nematode worm) (Rubin, 2001). (Throughout this article *homologous* means sequences that are so similar that the similarities cannot be due to chance but are the result of common ancestry.) However, even though exons and protein domains are shared by widely diverse species they often are present in novel combinations and arrangements in different organisms.

For instance, another vertebrate that has been sequenced is the puffer fish. The puffer fish was chosen because it has the smallest vertebrate genome so far discovered and as such is considered to be most similar to the common ancestor of the vertebrates. Its genome is one-seventh the size of the human genome. However, puffer fish appear to have virtually all the exons that are present in humans. One reason for the difference in genome size is that the lineage leading to hu-

FIGURE 1. The Three Domains (Archaea, Bacteria, and Eucarya) That Contain All Forms of Life Found on Earth.

Phylogenetic relationships are based on homology of ribosomal RNA sequences. The distance between groups is proportional to their genetic differences and evolutionary relatedness (after Woese, 1998, and McKane and Kandel, 1996).

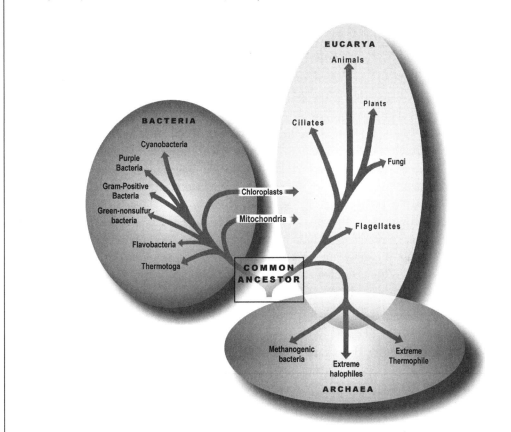

mans contains a great deal of duplication, both of genes and whole chromosome segments. Consequently, some of the puffer fish exons have been duplicated and reduplicated in humans and in many cases rearranged into new combinations. In this way it is possible for humans to have twice as many genes as puffer fish with the same number of exons.

Based on these observations from comparative genomics, vertebrate evolution has required the invention of very few new protein domains (Rubin, 2001). Thus, one aspect of evolutionary change involves making new genes by rearranging functional domains into novel combinations. This process is called exon shuffling. From comparative genomics we have discovered that making new genes by exon shuffling is a very important source of genetic variation, beyond mutation and recombination, upon which natural selection can act. The mechanisms by which exon shuffling occurs are under active investigation (Long, 2001).

Evolution and Creationism

Creationists do not all fit into one category. In the past creationists contended that species were created by God and did not change. This "fixity of species" concept was universal prior to Darwin and Wallace and is still held by some creationists. However, many creationists today concede, in the face of overwhelming evidence, that species change. Present day examples of antibiotic resistance in bacteria and insecticide resistance in insects alone are enough to establish that fact. But creationists continue to insist that such changes can only occur within narrow limits. Some insist that such changes can only result in adaptive changes within the species. Others are more liberal, allowing for natural selection acting on genetic variation, to produce new species but only within a "kind."

Then, within the last several years a more sophisticated version of creationism has emerged, "intelligent design" (ID). The ID proponents prefer not to be called creationists, not because they do not believe in a creator, but because they consider their evidence against evolution to be entirely scientific rather than faith based, as is the case with more traditional varieties of creationism. The "scientific evidence" adduced by the ID proponents consists of observing complexity at the cellular, molecular, and biochemical levels and asserting that such complexity could not have been produced by natural selection acting on random genetic variation (i.e., by microevolution). Thus, by their assertion cells are too complex to have arisen from preformed organic molecules, and different "kinds" of organisms could not have evolved from a common ancestor because the differences between them are too complex. They contend that microevolution is not up to the task of creating cellular complexity or new kinds of organisms—these can only result from design.

However, evolutionists argue that organisms sharing common traits are phylogenetically related through a common ancestor and have not been specially created, assertions and contentions by creationists to the contrary notwithstanding. The evidence from morphology, anatomy, embryology, physiology, genetics, and vestigial structures and organs has been sufficient to convince nearly all scientists that different kinds of organisms are biologically related and so have evolved from a common ancestor. The molecular evidence amassed in recent years and discussed in this article provides even more support for evolutionary relatedness.

At the level of whole genes, 60 percent of the human genes that encode proteins are homologous to genes from other organisms. Looking from the perspective of other organisms, 46 percent of yeast (*Saachromyces cervisiae*) genes, 43 percent of worm (*Caenorhabditis elegans*) genes, and 61 percent of fruit fly (*Drosophila melanogaster*) genes show sequence similarity with human genes (Lander et al., 2001). In summary, the high degree of conservation of both genes and exons among widely diverse organisms from all three phylogenetic domains is strong evidence for their common ancestry.

Nonfunctional Sequences

Perhaps even stronger evidence for relatedness among diverse organisms than similarity among functional genes comes from similarity in DNA sequences that have no function. One category of nonfunctional sequence is the pseudogene. There are two kinds of pseudogenes. One kind arises from a gene duplication followed by mutations to stop codons in one of the duplicates with the other retaining the original function. The other kind of pseudogene are recognized because they, like messenger RNA, have a poly A tail sequence and lack a promoter sequence and introns. Without a promoter they cannot be transcribed. These are called processed pseudogenes because they evidently arise through reverse transcription of messenger RNA into double-stranded DNA, which then is incorporated into the genome.

To date, 2909 processed pseudogenes have been recognized in the human genome (Lander et al., 2001). The following pseudogene example supports relatedness and common ancestry. Figure 2A shows the beta hemoglobin gene cluster for two of the most distantly related primates—humans and *Galago* (Bush Baby). The cluster consists of beta (β), delta (δ), pseudogene eta ($\Psi\eta$), gamma (γ), and epsilon (ε) genes. By sequence similarity, these genes are seen to have arisen by a series of duplications followed by mutational divergence, giving rise to their present functional differences that are expressed at different stages of embryonic, fetal, and adult development. The times at which the various hemoglobin duplications occurred are shown in Figure 2B. Of particular evolutionary interest is the conserved presence and conserved position of the functionless $\Psi\eta$ in all primate species (Goodman, 1999). By far the most plausible explanation for this observation is that these species are all related and that the $\Psi\eta$ was present in the common ancestor. The standard creationist explanation for similarities among functional genes is that since genes control traits, organisms with similar traits would of course have similar genes. This explanation does not work for pseudogenes.

There is no plausible intelligent design explanation for the occurrence of this particular functionless gene in every primate species, and the creationist argument that this is an example of the creator expressing artistic creativity rings hollow.

Another category of functionless DNA involves long interspersed nucleotide elements (LINEs). These are not entirely noncoding but they are nonfunctional as far as the individual organism is concerned. In the human genome there are three families of LINES: LINE 1, LINE 2, and LINE 3. There are 516 000 copies of LINE 1, 315 000 of LINE 2, and 37 000 of LINE 3 (Lander et al. 2001). Lines are now recognized to be retrotransposons, which are generally considered to be defunct retroviruses. Retroviruses are RNA viruses that—when they enter cells—cause their RNA to be converted into a double-stranded DNA molecule that is inserted into the host's DNA from which it controls the production of new viruses. LINEs retain the viral genes that allow their transcripts to be reverse transcribed into DNA and inserted into chromosomes, but have lost the genes for making coat proteins and escaping from the cell. In this way they transpose themselves into new chromosomal locations in a retrovirallike manner and so are called retrotransposons. This explains the thousands of copies.

SECTION III — THE SCIENCE TEACHER'S PERSPECTIVE

FIGURE 2A — Beta Hemoglobin Gene Cluster as It Occurs in Primates.

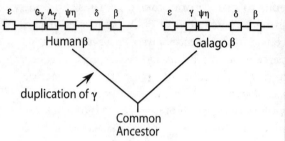

The pseudogene eta ($\psi\eta$) shown here is also present in all other primates indicating their relatedness. (Note: Some textbook figures do not show the pseudogene.) Humans have acquired a duplication of the gamma gene, one designated with an A and the other G. These stand for adenine and guanine, the single nucleotide by which they differ. This duplication is present in all the simians (apes and monkeys) as well as in humans, but is not present in other primates. It is extremely unlikely that there is a physiological need for two copies of the gene in this group of primates but none of the others. Thus we conclude that monkeys, apes, and humans share this duplication because it arose in the common ancestor.

FIGURE 2B — Evolutionary History of the Duplications in the Hemoglobin Family.

(Numbers in parentheses are the estimated number of nucleotide replacements in the branches shown.)

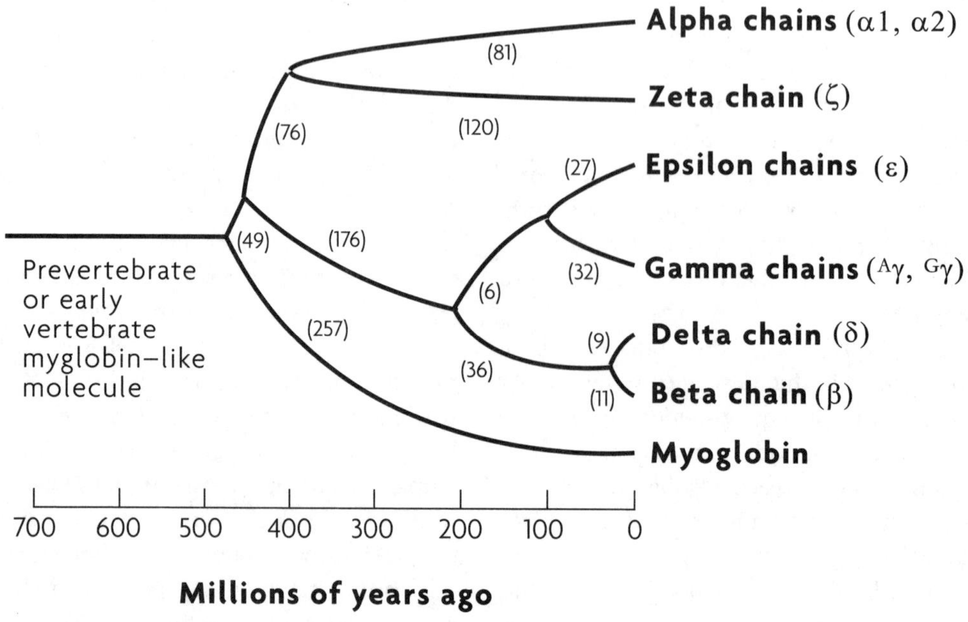

Prevertebrate or early vertebrate myglobin–like molecule

- Alpha chains ($\alpha 1, \alpha 2$) — (81)
- Zeta chain (ζ) — (120)
- Epsilon chains (ϵ) — (27)
- Gamma chains ($^A\gamma, ^G\gamma$) — (32)
- Delta chain (δ) — (9)
- Beta chain (β) — (11)
- Myoglobin

Branch values: (76), (49), (176), (6), (257), (36)

Millions of years ago: 700 600 500 400 300 200 100 0

Strickberger, M.W. *Evolution*, 2000. Sudbury, Mass.: Jones and Bartlett Publishers. www.jbpub.com. Reprinted with permission.

If LINEs in different species are homologous (by sequence similarity) this is very strong evidence that the species share a common ancestor in which the LINE first became established. Otherwise the virus would have to have become defunct independently in each of the species, which for even two species is considered to be very unlikely. The LINE 1 sequence is present not only in humans but in every species of mammal examined to date (Smit et al., 1995). Not only does this show indisputably that all mammals are related, a point that creationists deny, but also that LINE 1 is very old.

In addition to LINEs in genomes of various organisms, there are short interspersed nucleotide elements (SINEs), which also can move around and are present in multiple copies. There are three families of SINEs in the human genome, collectively consisting of 1 558 000 copies (Lander et al., 2001).

Let us consider an example involving recent interest in the relationship of whales to other mammals. There are several specific LINEs and SINEs that are shared by whales and artiodactyls (the order of cloven-hoofed mammals), which are not present in other mammals. The likelihood that even one of these shared sequences arose independently in all of these species is virtually zero. Consequently, the conclusion is inescapable that all artiodactyls species are related to each other and also to whales (Nikaido, Rooney, and Okada, 1999). That is, they share a common ancestor from whom they inherited the LINEs and SINEs that they share. Moreover, once a LINE or SINE is established in a specific location it does not leave. Thus if a particular LINE or SINE is found in identical locations in each of two species that also is an indication of common ancestry. One SINE, Chr-1, is found in exactly the same position, to the nucleotide pair, in four different genes of whales and hippopotami. Chr-1 is not found in any of these locations in any other artiodactyl species. Thus whales and hippopotami clearly are each other's closest living relatives (Nikaido, Rooney, and Okada, 1999), necessitating the establishment of a new suborder, Cetartiodactyla.

The presence of conserved, but nonfunctional, sequences in a variety of different organisms is more consistent with an ongoing evolutionary process than creationist arguments invoking intelligent design or artistic creativity.

Universality of Triplet Code and Translation Process

Protein synthesis is a highly conserved fundamental life process and is responsible for most metabolic activities and phenotypes in all organisms. All living organisms synthesize proteins; it is a universal characteristic of life. Moreover, all organisms do it the same way, using messenger RNA, transfer RNA, ribosomes, and the same 20 amino acids. The triplet genetic code is virtually universal. The few exceptions, such as in a few ciliates and the mitochondria of some organisms, are all easily understood as single step changes from the standard code. Also, the processes of transcription and translation for protein synthesis are highly conserved among all forms of life, indicating that very little change has taken place since the time of the common ancestor over three billion years ago. Although the molecules involved in transcription and translation have undergone some changes in nucleotide and amino acid sequence, their function has not changed.

Differences in ribosomal RNA sequences determined the three domains, while they continue to carry out the same function in the small ribosomal subunit. The universality of both the genetic code and the method of protein synthesis is another strong argument for the relatedness of all organisms, because such a high degree of complex metabolic similarity did not evolve millions of times independently. Similarity and conservation of complex processes exist not just at the metabolic level, but also at every level from the molecular to the morphological. What is the explanation for complex similarities? The argument for evolution, based on genetic descent with modification, is powerful and persuasive.

BIOCHEMICAL EVIDENCE OF EVOLUTIONARY RELATIONSHIPS

Metabolic Pathways

Biochemists and geneticists have accumulated experimental evidence since the 1930s that all organisms contain interdependent biochemical and metabolic pathways that fail to work if one component becomes defective. Creationists have pounced upon this fact to assert that all of the steps in a metabolic pathway must therefore necessarily have arisen simultaneously and because that has such a low probability of occurrence, species must have been created (Behe, 1996).

Evolutionists, on the other hand, argue that complex pathways have been built up by adding one step at a time. One line of evidence for stepwise addition is the fact that genes for the enzymes in a pathway are frequently so similar in sequence that they clearly have arisen as a series of gene duplications from one original gene, and this would necessarily have occurred in stepwise fashion (Chothia et al., 2003).

The basic chemical composition of cells (i.e., carbon, hydrogen, oxygen, nitrogen, phosphorus, and sulfur) and many aspects of cellular metabolism are also highly conserved. For instance, virtually all present-day organisms, both aerobic and anaerobic, carry out glycolysis, the conversion of 6-carbon glucose molecules to two 3-carbon pyruvic acid molecules. This is because the glycolytic metabolism of glucose to pyruvic acid has been conserved in all three domains from the time of the common ancestor.

The Krebs Cycle and its associated electron transport system are present in all aerobic eucaryotes because the reactions are carried out by mitochondria. Since virtually all biologists agree that mitochondria arose through bacterial endosymbiosis, this event must have occurred in the common ancestor of all present day eucaryotes. The creationists would have to deny that mitochondria in all eucaryotes, both plants and animals, and chloroplasts in plants arose by endosymbiosis.

At the metabolic level, just as at the levels of morphology, anatomy, and development, the same complex traits require similar sets of genes to produce them. Genes are passed from one generation to the next, therefore the most plausible explanation for complex genetic similarities is common ancestry.

Genetic Variability of Proteins

Genes and their encoded proteins that perform the same metabolic functions in different organisms are similar but in most cases not identical with respect to their DNA and protein sequences. For example, the cytochrome C protein that performs the same electron transport function in horse and cow mitochondria is very similar, but not identical. In fact, comparing cytochrome C among 60 different species examined revealed that only 27 amino acid residues are identical in all 60 while more than 60 residues differ among them (the latter number is not exact because there are slight differences in the length of the molecule in some species).

However, the degree of similarity among amino acid sequences in cytochrome C corresponds closely to the phylogenetic relationships based on other criteria. That is, mammalian sequences are more similar to each other than to any reptilian sequence and vice versa, and so on. Similar patterns exist for other proteins that have been compared among species. For instance, the human hemoglobin alpha chain differs from that of sharks by 79 out of 141 amino acids, from a

Topic: Proteins
Go to: *www.scilinks.org*
Code: EIP12

bony fish (carp) by 68, an amphibian (newt) by 62, a bird (chicken) by 35, a horse by 18, and a chimpanzee by 0 (Figure 3).

Mammals are more similar to each other than any mammal is to a bird. There is no plausible intelligent design explanation for these enormous sequence differences at both the protein and nucleotide levels corresponding to degrees of relatedness based on other criteria. But evolution explains them perfectly. Such genetic and biochemical changes have taken place over millions of years of evolutionary time by a combination of selective improvements and fixation of neutral mutations that have no effect on protein function.

On the other hand, despite the enormous potential for flexibility, comparison of the amino acid sequences of nine different proteins in humans and chimpanzees (including hemoglobin and cytochrome C) reveals a total of only five amino acid differences (Strickberger, 2000). This is very strong evidence that the two species are very closely related and are each other's closest relative.

GENETICS OF EVOLUTION

With Mendelism it became clear that genes determine the characteristics of individual organisms and that they occasionally mutate. This much knowledge was enough to illuminate the relationship between mutation and natural selection; between genetics and evolution. Indeed, Mendel's laws of genetic inheritance, rather than some fossil intermediate between two presently existing species, were Darwin's missing link. The understanding that genetic mutations produce the genetic variation in natural populations upon which natural selection acts was consolidated in

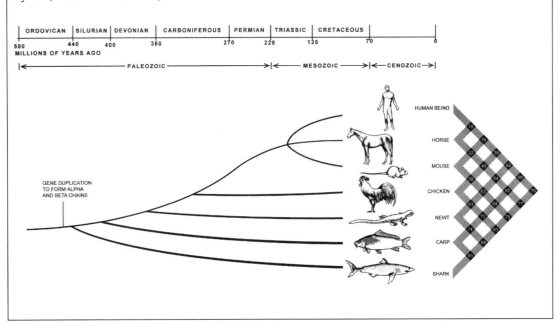

FIGURE 3: The Number of Amino Acid Differences in the Alpha Hemoglobin Chain between the Representative Species.

The number of amino acid differences between any two species is read at the intersection of the diagonals. For instance, shark differs from carp by 85, from newt by 84, and from chicken by 83 (from Kimura, 1979).

the 1930s (the neo-Darwinian synthesis). Mutation and natural selection were finally seen as complementary factors, rather than competing mechanisms in the origin of species. We could see how, in the words of the influential evolutionary biologist Theodosius Dobzhansky (1951), "evolution is a change in the genetic composition of populations."

Out of the neo-Darwinian synthesis grew the field of population genetics. Geneticists strove to determine how much genetic variation existed in natural populations and what factors other than mutation and natural selection, such as migration, geographic isolation, and chance (genetic drift), might interact with natural selection to change gene frequencies. Mathematical formulae were developed to calculate the effects of these various factors on gene frequencies in populations, and various modes of speciation were investigated. These studies continue today and have been joined by ecologists who study specific interactions between organisms and their environments to determine environmental effects on gene frequencies. All of this began before we learned that genes specify proteins, or that genes were DNA molecules.

Gene Duplication and Evolution

Another question that arose was the origin of new genes. Obviously more complex organisms need more genetic information than simple organisms, although we were surprised to learn that humans have only slightly more than twice as many genes as *Drosophila* (Venter et al. 2001). If complex organisms evolved from simple ones where did the new genes come from? It was hypothesized from the outset that on rare occasions a small section of chromosome, along with the genes it contains, is duplicated. The mechanism or mechanisms for this are not entirely clear, perhaps by unequal crossing over, but their existence is indisputable. After duplication, one copy of a gene can maintain the original function while the other is free to mutate to a different function.

There was abundant evidence from *Drosophila* for small duplications, and as shown in Figure 2A (p. 84), the occurrence of small duplications can be seen directly in the globin gene families of vertebrates. Moreover if we compare the relative differences among the various globin genes, as was done for humans, it is possible to establish the evolutionary history of the globin family. From these differences we can deduce that the alpha chain gene arose from a duplication of the myoglobin gene and the beta chain gene from a duplication of the alpha chain gene, and approximately how long ago the duplications occurred (Figure 2B, p. 84). Additional duplications then occurred in the alpha and beta gene clusters. The alpha cluster now consists of alpha 1, alpha 2, and zeta, and the beta cluster consists of beta, delta, gamma, and epsilon. If a group of animals, such as the primates in Figure 2A, share a complex gene cluster that arose by a series of duplications it provides overwhelming evidence that the cluster existed in a common ancestor and that the species in the group are biologically related rather than produced by an act of special creation.

Exon Shuffling and Evolution

We have already indicated that exon shuffling as a method for making new genes is a reality. But let us consider a specific example that illustrates both exon shuffling and serial duplication. It involves the genes encoding the enzymes for the mammalian blood-clotting cascade, which, incidentally, is one of the creationist's favorite examples of a system that is too complex to have evolved (Behe, 1996). By sequence analysis it has been shown that these genes contain exons homologous to exons in the trypsin gene and others homologous to those in the gene for epider-

Topic: Genetic Variation
Go to: *www.scilinks.org*
Code: EIP13

mal growth factor (Doolittle, 1993). Each of the six enzymes in the cascade is, like trypsin, produced in an inactive form that is activated by having a segment of its amino acids removed. Each active enzyme in the cascade is, again like trypsin, a serine protease that functions to activate the next enzyme in the cascade.

Finally, prothrombin is converted to thrombin, which converts fibrinogen into fibrin and the clot forms. For the proenzymes to be cleaved they must be bound to cell membranes in the damaged tissue. This takes place because there is a membrane receptor called "tissue factor" that binds to the epidermal growth factor end of the proenzymes. And just as the genes for the enzymes in the cascade are, by sequence similarity, homologous to genes with other but related functions, the gene for tissue factor is homologous to the gene for epidermal growth factor receptor. The gene for fibrinogen is homologous to a gene for actin, a widespread contractile molecule, best known for its presence in muscles. From these observations alone the blood-clotting cascade looks more like a case of evolutionary tinkering, working with what is already available, than intelligent design. When one takes into consideration the high degree of sequence similarity between the enzymes in the cascade, it seems obvious that the first blood-clotting gene arose by exon shuffling, and subsequent genes of the cascade arose from simple gene duplication followed in each case by mutation to new, but related, functions.

Genome Duplication and Evolution

In order for exon shuffling to work there must be duplicate sets of genes if the original functions are to be retained. The human genome project has shown that the human genome is extensively duplicated, so extensively that it is debated whether the duplications arose from successive episodes of polyploidy (i.e., the doubling of whole sets of chromosomes) or by some other mechanism.

Gene duplication and selection for novel function likely has occurred also within bacteria. For example, sequence data reveal that the *Escherichia coli (E. coli)* genome is 4.6 million base pairs, the *Pseudomonas aeruginosa* genome is 6.2 million base pairs, and the *Streptomyces coelicolor* genome is 8.6 million base pairs. While all of these organisms are procaryotic, the range in genome size suggests that habitat diversity is selected for larger genomes with increased functional diversity. Also, the *E. coli* genome evolved to be 90 percent coding sequence, whereas more than 90 percent of the human genome is noncoding. An additional major difference is that the coding region of a procaryotic gene is a continuous exon but the coding region of a higher eucaryotic gene is segmented with introns and exons. The promoter region is also very different between procaryotic and eucaryotic genes, as demanded by the greater regulatory sophistication required by the metabolic and developmental complexity of eucaryotes.

Teaching Evolution Directly

In this summary we have emphasized relatedness among very different kinds of organisms rather than the fact that species change. Most creationists now concede that species change, even to the extent of giving rise to new species. However, creationists continue to insist that these changes are "microevolutionary," leading only to modifications within a "kind" and to nothing fundamentally new. All of the molecular evidence to date indicates that presently existing organisms are related and so must necessarily have arisen from a common ancestor through a process of evolution.

Are the fundamental similarities and conserved relationships discussed in this summary due to divine intervention, or do they reflect an evolutionary relatedness? The biochemical, genetic, and functional relatedness observed today among all forms of life support biological evolution. Our experience has been that discussing these lines of evidence in our workshops gives teachers more confidence to teach evolution directly, rather than evasively or not at all.

ACKNOWLEDGMENT
The authors thank Richard Edelmann of Miami University for excellent assistance with the figures.

REFERENCES
Asfaw, B., W.H. Gilbert, Y. Beyene, W.K. Hart, P.R. Renine, G. Woldegabriel, E.S. Vrba, and T.D. White. 2002. Remains of Homo erectus from Bouri, Middle Awash, Ethiopia. *Nature* 416:317–320.

Behe, M. 1996. *Darwin's Black Box: The Biochemical Challenge to Evolution*. New York: The Free Press.

Chothia, C., J. Gough, C. Vogel, and S.A. Techmann. 2003. Evolution of the protein repertoire. *Science* 300:1701–1703.

Dobzhansky, T. 1951. *Genetics and the Origin of Species*. New York: Columbia University Press.

Doolittle, R.E. 1993. The evolution of vertebrate blood coagulation: A case of yin and yang. *Thrombosis and Haemostasis* 70:24–28.

Glossary of terms

Comparative genomics: The explosive advances in sequencing technology have made it possible to sequence the genomes, and parts of genomes, of a number of species, including humans. Comparative genomics involves analyzing and comparing these sequences—it has been a treasure trove of evidence supporting evolution.

Duplications: Duplications of DNA range in size from a few nucleotides, to chromosomal segments, to entire chromosomes to entire genomes. Duplications are the source of DNA for new genes and abound in every genome thus far studied.

Exons: The portions of genes that actually code for amino acids in proteins. Typically, exons code for functional domains in proteins (see introns and functional domains).

Exon shuffling: This process recombines exons from existing genes into novel combinations in new genes. The mechanisms for exon shuffling are not well understood but are under active investigation.

Functional domain: Proteins typically consist of regions, or domains, exhibiting quasi-independent functions. For instance, a single protein might have a cytoplasmic domain anchoring it to the cytoskeleton (microfiber or microfilament), a transmembrane domain, and an extracellular domain that binds a hormone. Typically, a separate exon codes for each functional domain.

Introns: DNA sequences interspersed among the exons of genes. These sequences are transcribed, with the exons, into premessenger RNA, but are removed as the exons are joined together to form the final messenger RNA. Introns are not well understood but may play a role in alternate splicing, a mechanism for obtaining different messenger RNA molecules from the same gene. Nearly all eucaryotic genes have introns varying in number from one to many. They are present, but less common, in Archea, and virtually absent from bacteria.

LINEs: Long interspersed nuclear elements (LINEs) typically are present in many thousands of copies in most eucaryotic organisms. LINEs are perceived to contain genes that facilitate their movement to new

Goodman, M. 1999. The genomic record of humankind's evolutionary roots. *American Journal of Human Genetics* 64:31–39.

Hardison, R., J.L. Slighton, D.L. Gumucio, M. Goodman, N. Stojanovic, and W. Miller. 1997. Locus control regions of mammalian beta-globin gene clusters: Combining phylogenetic analyses and experimental results to gain functional insights. *Gene* 205:73–94.

Horiike, T., K. Hamada, S. Kayana, and T. Shinozawa. 2001. Origin of eukaryotic cell nuclei by symbiosis of Archea in Bacteria is revealed by homology hit analysis. *Nature Cell Biology* 3:210–214.

Kimura, M. 1979. The neutral theory of molecular evolution. *Scientific American* 241(5): 94–104.

Lander, E.S., et al. 2001. Initial sequencing and analysis of the human genome. *Nature* 409:860–921.

Long, M. 2001. Evolution of novel genes. *Current Opinion in Genetics and Development* 11:673–680.

Martin, W., and M. Muller. 1998. The hydrogen hypothesis for the first eucaryote. *Nature* 392:37–41.

---

chromosomal locations via a transcribed RNA intermediate. They are now called retrotransposons or retroposons to reflect the view that they are defunct retroviruses (see SINEs).

Nucleotides: Molecular subunits of DNA and RNA.

Promoter region: The region of a gene that serves as a recognition site for RNA polymerase to initiate transcription. Other proteins called transcription factors interact with the promoter region and RNA polymerase to help regulate transcription. In eucaryotes such as humans there may be as many as 70 or 80 different proteins interacting with the promoter sequence at different times of development. The promoter region itself is not transcribed. The length of the promotor region varies from approximately 50 nucleotides in procaryotic genes to several thousand in eucaryotic genes.

Pseudogenes: Genes that do not produce functional proteins. Processed pseudogenes, so called, are derived from messenger RNA and so are not transcribed because they lack promoter sequences. Other pseudogenes arise by gene duplication followed by mutation to stop codons in one of the duplicates. This results in protein fragments.

Serine proteases: A class of protein digesting enzymes with a characteristic active site that includes the amino acid serine. The blood-clotting enzymes are serine proteases.

SINEs: Short interspersed nuclear elements (SINEs) in eucaryotic organisms. These also occur in many thousands of copies and like LINEs they have some of the telltale characteristics of movable elements. However, they lack the genes that are present in LINEs and perhaps depend on LINE genes for their movement.

16S and 18S: The sedimentation coefficients used to describe the relative size of the RNA molecule in the small ribosomal subunit of bacteria (16S) and eucaryotes (18S).

McKane, L., and J. Kandel. 1996. *Microbiology*. New York: McGraw-Hill.

Nikaido, M., A.P. Rooney, and N. Okada. 1999. Phylogenetic relationships among cetartiodactyls based on short and long interspersed elements: Hippopotamuses are the closest extant relatives of whales. *Proceedings of the National Academy of Sciences* 96(18): 10261–10266.

Rubin, G.M. 2001. Comparing species. *Nature* 409:820–821.

Smit, A., C. Toth, A. Riggs, and J. Jurga. 1995. Ancestral mammalian-wide subfamilies LINE 1 repetitive sequences. *Journal of Molecular Biology* 246:401–417.

Strickberger, M.W. 2000. *Evolution*. 3rd ed. Sudbury, Mass.: Jones and Bartlett.

Venter, C., et al. 2001. The sequence of the human genome. *Science* 291:1304–1351.

Woese, C.R. 1998. The universal ancestor. *Proceedings of the National Academy of Sciences* 95:6854–6859.

APPENDIX

List of Contributors

Francisco J. Ayala, author of "Arguing for Evolution," is the Donald Bren Professor of Biological Sciences, Department of Ecology and Evolutionary Biology, at University of California, Irvine.

DeWayne A. Backhus, author of "It's Not Just a Theory," is Chair of the Departments of Physical Sciences at Emporia State University in Emporia, Kansas.

J.K. Bhattacharjee, co-author of "A Guide to Teaching Evolution," is a professor of microbiology at Miami University in Oxford, Ohio.

Rodger W. Bybee, author of "Evolution and the Nature of Science" and "Evolution: Don't Debate, Educate," is Executive Director of the Biological Sciences Curriculum Study in Colorado Springs, Colorado.

Emily Case, author of "Symbiosis: An Evolutionary Innovator," is a science teacher at Smith Academy in Hatfield, Massachusetts.

G. Brent Dalrymple, author of "Evidence for Evolution," is Professor Emeritus at Oregon State University in Corvallis, Oregon.

Anthony V. DeFina, author of "Investigating Island Evolution," is the science department chair at Wayne Hills High School in Wayne, New Jersey.

George Glasson, co-author of "Comparing Common Origins," is an associate professor of science education, Department of Teaching and Learning, at Virginia Tech in Blacksburg, Virginia.

Thomas G. Gregg, co-author of "A Guide to Teaching Evolution," is a professor of evolutionary and molecular biology in the department of zoology at Miami University in Oxford, Ohio.

Gary R. Janssen, co-author of "A Guide to Teaching Evolution," is an associate professor of microbiology at Miami University in Oxford, Ohio.

John McLaughlin, co-author of "Comparing Common Origins," is a biology teacher at Lord Botetourt High School in Daleville, Virginia.

Jill C. McNew, co-author of "Attitudes toward Evolution," is a science teacher at Street School, Inc., in Tulsa, Oklahoma.

APPENDIX

John A. Moore (deceased), author of "Thought Patterns in Science and Creationism," was one of the founders of Biological Sciences Curriculum Study (BSCS), and distinguished biologist and Professor Emeritus of biology at the University of California, Riverside.

Randy Moore, author of "Do Standards Matter?," is professor of biology at University of Minnesota–General College in Minneapolis, Minnesota.

Susan Stone Plati, author of "Searching for the Perfect Lesson," is a biology teacher at Brookline High School in Brookline, Massachusetts.

John R. Staver, author of "Evolution and Intelligent Design," is professor of science education and director of the Center for Science Education at Kansas State University in Manhattan, Kansas.

Jeffrey Weld, co-author of "Attitudes toward Evolution," is an associate professor of science education, School of Curriculum and Educational Leadership, at Oklahoma State University in Stillwater, Oklahoma.

Index

(Page numbers printed in **boldface** type refer to figures.)

A

A Kansan's Guide to Science, 38–39
AAAS. *See* American Association for the Advancement of Science
Aguillard v. Treen, 5
Alberts, Bruce, xvi
American Association for the Advancement of Science (AAAS), xvi, xix
    Benchmarks for Science Literacy, x, xvi, xxi, 50
    position on teaching of creationism, 50
    Science for All Americans, xvi, 51
American Institute of Biological Sciences, xvi
Aristotle, 45
Attitudes toward evolution, ix–x, 26–27, 49–55
    comparison of subpopulation trends in, 53–55
    current views of science education community, 50–51
    methodology for study of, 51–52
    overall trends in, 52–53, **53**

B

Benchmarks for Science Literacy, x, xvi, xxi, 50
Biblical accounts of creation, xxi, 3, 10
Big Bang, 7, 11
Biochemical evidence for evolution, 86–87
    genetic variability of proteins, 86–87, **87**
    metabolic pathways, 86
Bioinformatics, 68
Biological Sciences Curriculum Study (BSCS), xii, 9
    BSCS Biology: A Human Approach, xii
Biotechnology, 67–72
    definition of, 67
    protein electrophoresis lab, **68**, 68–72
"Black box" activities, 40, 41
Blood-clotting cascade, 88–89
Bobbit, John, 39

C

Carnegie Foundation for the Advancement of Teaching, 39
Cladograms, 71–72
Cohen, I. B., xv
Comparative genomics, 80–83, 90
Copernicus, xv
Creation science, xxi–xxii, 5–6, 14, 50
Creationism, x, xii–xiii, xvi, xxi–xxii, 5, 82
    avoiding debates about, 30–32, 34
    definition of, xxi
    "fixity of species" concept, 82
    increasing influence of, 9
    judicial decisions regarding teaching of, xxii, 5
    lack of evidence for, 11
    P version of, 10
    religion and science, 2–3
    science teachers' opinions about, 26–27, 49–55, **53** (*See also* Attitudes toward evolution)
    special, xxi
    thought patterns in science and, 9–14
Cremin, Lawrence, 39
Cytochrome c, 86

D

Darwin, Charles
    exploration of Galapagos Islands, 58–59
    theory of natural selection, **xiv,** 23, 64
Darwin on Trial, 45
deBroglie, Louis, 44
Definition of science, xv
Dembski, William, 43
Dewey, John, 39
DNA, 67–68
    nonfunctional sequences of, 83–85, **84**
    universality of structure of, 80

INDEX

Dobzhansky, Theodosius, 88
Duplications, 82, 90
    gene, 83, **84,** 88
    genome, 89
Duschl, Richard, 38

E
Earth, dating of, 6–7
Edwards v. Aguillard, xxii, 5
18S ribosomal RNA, 80, 91
Eukaryotic genes, 80
Evidence for evolution, 5–8
    acceptance of, 11–12
    the Big Bang, 7, 11
    biochemical, 86–87
    vs. creationism, 9–14, 82
    dating the Earth, 6–7
    genetic, 87–89
    molecular, 79–86
    star dates, 7–8
    teaching of, 14
    validation of, 10–11
Evolution Project, 65
Evolutionary theory, xii–xiii, 1–2
    attitudes toward, ix–x, 26–27, 49–55 (*See also* Attitudes toward evolution)
    Darwin's theory of natural selection, **xiv,** 23, 64
    evidence for, 5–8
    formulation and development of, xiii
    intelligent design and, 3–4, 43–48, 82
    in National Science Education Standards, **16,** 16–17
    presenting evidential basis for, 14
    teaching of (*See* Teaching evolution)
    thought patterns in creationism and, 9–14
    as unifying concept, xx–xxi
Exon shuffling, 82, 88–89, 90
Exons, 80–82, 90
Expansion of the universe, 7

F
"Fixity of species" concept, 82
Functional domain, 81, 90

G
Galapagos Islands evolution study, 57–62
    classroom exploration phase of, 58–59
    elaboration phase of, 60–61
    engaging islands, 58
    rubric for scoring essays on, **61**
    species notes used during elaboration phase of, **60**
    species notes with suggestions for student exploration, **59**
    student evaluation phase of, 61–62
    student explanation of evolution, 59–60
Galileo, xv
Genetic mutations, 76, 82, 87–88
Genetics of evolution, 87–89
    exon shuffling, 88–89
    gene duplication, 88
    genome duplication, 89
    glossary of terms for, 90–91
Globin gene family, **84,** 86–87, 88
Good Science, Bad Science: Teaching Evolution in the States, 25
Gray, Thomas, 12

H
Hidden curriculum, 39
Hubble, Edwin, 7
Human genome, 80
    duplications in, 89
    pseudogenes in, 83

I
Inquiry-based activities, 32–34, 41
Institute for Creation Research, 52
Intelligent design (ID), 3–4, 43–48, 82
    classroom controversy and, 47–48
    establishing new knowledge, 44–45
    evidence for, 45–46
    evolutionary theory and, 3–4, 44
    reasons ID is not a scientific theory, 46–47
Introns, 80, 90

INDEX

J
Jackson, P. W., 39
Johnson, Phillip, 45
Judicial decisions, xxii, 5

K
Kennedy, Don, xvi
Krebs Cycle, 86

L
Lane, Leslie, 45–46
Legal issues, xxii, 5
Lerner, Lawrence, xvi, 25–26
Lessons on evolution, 63–66
    engaging all students in, 64–65
    Galapagos Islands study, 57–62
    multimedia presentations, 65–66
    protein electrophoresis lab, **68,** 68–72
    symbiosis, 73–77
Lichens as example of symbiosis, 74, **74,** 75
Long interspersed nuclear elements (LINEs), 83–85, 90–91
Lunar rocks, dating of, 6–7

M
Mayr, Ernst, xiii, xv
McLean v. Arkansas Board of Education, xxii, 5
McNew, Jill, ix–x
Mendel, Gregor, 45
Mendelian inheritance, 87
Metabolic pathways, 86
Meteorites, dating of, 7
Milky Way Galaxy, dating of, 6, 7–8
Miller, Ken, xvi
Modern Biology, 45
Molecular evidence for evolution, 79–86
    nonfunctional DNA sequences, 83–85
    similarities among diverse organisms, 80–83
    universality of DNA structure, 80
    universality of triplet code and translation process, 85–86
Moore, John A., xv
Moore, Randy, xvi
Muller, Hermann, 27
Multimedia presentations, 65–66

N
NABT (National Association of Biology Teachers), xvi, 27, 52
National Academy of Sciences (NAS), xvi, 31
    position on "creation science," 5–6
    Science and Creationism, 23, 39
    Teaching about Evolution and the Nature of Science, xv, xxi, 23, 39
National Association of Biology Teachers (NABT), xvi, 27, 52
National Center for Biotechnology Information (NCBI), 69
National Center for Science Education, xvi, 30
National Earth Science Teachers Association (NESTA), xvi
National Research Council, 15, 23
National Science Education Standards (NSES), ix, x, xv–xvi, xxi, 15–23, 38
    development and usage of, 21–23, 45
    for teaching about nature of science, 17–21, 34
        grades K–4, 18–19
        grades 5–8, 19, **21**
        grades 9–12, 19–21, **22, 33**
    for teaching evolution, **16,** 16–17, 50, 67
        grades K–4, 16–17
        grades 5–8, 17, **18**
        grades 9–12, 17, **19, 20**
National Science Teachers Association (NSTA), xvi, 31, 38
    Scope, Sequence, and Coordination Project, 50
    The Teaching of Evolution, x, xix–xxii, 50, 51
Natural selection, 3, 76
    Darwin's theory of, **xiv,** 23, 64
Nature of science
    "black box" activities and, 40, 41
    concepts and principles of, **33**
    inquiry, evolution and, 32–34
    standards for teaching about, 17–21, 34
    theory and, 37
NCBI (National Center for Biotechnology Information), 69
Neo-Darwinian synthesis, 88

INDEX

NESTA (National Earth Science Teachers Association), xvi
NSES. *See* National Science Education Standards
NSTA. *See* National Science Teachers Association
Nucleotides, 80, 91

O

On the Origin of Species by Means of Natural Selection, 32, 64
One Long Argument, xiii, xv, 32

P

Paley, William, 45
Peloza v. Capistrano Unified School District, xxii
Phylogenetic trees, 68–72, **69**
Polyploidy, 89
Pope John Paul II, xxi, 3, 4
Population genetics, 88
Promoter region, 83, 91
Protein electrophoresis lab, **68**, 68–72
    equipment and kits for, 68–69
    phase one: engagement, 69–70
    phase two: exploration, **70**, 70–71
    phase three: explanation, 71, **71**
    phase four: extension, 71–72
    phase five: evaluation, 72, **72**
    resources for, 69, 71, 72
Proteins, 81
    genetic variability of, 86–87
    synthesis of, 85
Pseudogenes, 83, **84**, 91

R

Radiometric dating of rocks, 6
Religion and science, 2–3, 5–6. *See also* Creationism
Restructuring Science Education: The Importance of Theories and Their Development, 38
Retrotransposons, 83
Retroviruses, 83
Revolution in Science, xv
Ribosomal RNA sequences, 80, 85, 91, **92**

Riddle, Oscar, 27
Rocks, dating of, 6–7
Rutherford, Ernest, 44

S

Sagan, Carl, xv
Science, xvi
Science and Creationism, 23, 39
Science as a Way of Knowing, xv, 32
Science curriculum, 38–39
Science for All Americans, xvi, 51
Science teachers
    attitudes toward evolution, ix–x, 26, 49–55
    endorsement of creationism by, 26–27
    support for, xv–xvi
    teaching strategies of, 38–40, xii, xiii (*See also* Teaching evolution)
Scientific theories, xiii, xx, 37–41. *See also* Evolutionary theory
    in the classroom, 38
Scientific way of thinking, xi, xv, xx
    vs. creationism, 9–14
SciLinks
    comparing common origins, 68
    Edwin Hubble, 7
    explore Mendelian genetics, 64
    Galapagos Islands, 58
    Newton's Laws, 39
    observing traits molecularly, 80
    proteins, 86
    Rutherford model of atom, 44
    symbiosis, 74
    symbiosis of fungi, 75
Scopes trial, xii
Scott, Eugenie, xvi
Serine proteases, 91
Short interspersed nuclear elements (SINEs), 85, 91
16S ribosomal RNA, 80, 91
St. Augustine, 3, 4
Star dating, 7–8
State standards for teaching evolution, 21, 25–27, 50–51
    grading of, xvi, 25–26

INDEX

in Kansas, 1, 9, 37–38, 49
Supernovas, 8
Symbiosis, 73–77
    definition of, 73–74
    evolution and, 74–75
    lichens as example of, 74, **74**
    role in shaping the genetic landscape, **75,** 75–76, **76**
    science education standards and, 77

T

Teaching about Evolution and the Nature of Science, xv, xxi, 23, 39
Teaching evolution, 79–90, xiii–xv
    current views of science education community on, 50–51
    defensiveness about, xii
    judicial decisions regarding, xxii, 5
    lack of accountability for, 27
    lessons for, 63–66
    multimedia for, 65–66
    NSTA position statement on, x, xix–xxii
    opposition to, xxii, 1
    presenting evidential basis for, 14
    resources for, xv, 23
    science teachers' attitudes about, ix–x, 26–27, 49–55, **53**
    state standards for, xvi, 21, 25–27
    strategies for, xii, xiii
    textbooks and, xii
    using biotechnology for, 67–72
    views of professional organizations on, xvi, 27
Teaching science and technology, 13–14
Textbook coverage of evolution, xii
The American Biology Teacher, xvi, 27
The Process of Education, 32
The Science Teacher, ix, xvi
The Teaching of Evolution, x, xix–xxii, 50, 51
This Is Biology, xv
Thomson, J. J., 44
Thought patterns in science and creationism, 9–14
Transcription and translation, 85

Triplet genetic code, 85–86

W

Webster v. New Lennox School District #122, xxii
Weld, Jeffrey, ix–x

Y

Young-earth creationists, 5–6